T0323634

Internet of Things

Internet of Things

Integration and Security Challenges

Edited by

*S. Velliangiri, Sathish A. P. Kumar,
and P. Karthikeyan*

CRC Press
Taylor & Francis Group
Boca Raton London New York

CRC Press is an imprint of the
Taylor & Francis Group, an **informa** business

First edition published 2021
by CRC Press
6000 Broken Sound Parkway NW, Suite 300, Boca Raton, FL 33487-2742
and by CRC Press
2 Park Square, Milton Park, Abingdon, Oxon, OX14 4RN

Library of Congress Cataloging-in-Publication Data

Names: Velliangiri, S., editor. | Kumar, Sathish A. P., editor. |
Karthikeyan, P., 1981- editor.
Title: Internet of things : integration and security challenges / edited by
S. Velliangiri, Sathish A. P. Kumar, and P. Karthikeyan.
Other titles: Internet of things (CRC Press)
Description: First edition. | Boca Raton, FL : CRC Press/Taylor & Francis
Group, LLC, 2021. | Includes bibliographical references and index.
Identifiers: LCCN 2020032513 (print) | LCCN 2020032514 (ebook) | ISBN
9780367893873 (hardback) | ISBN 9781003032441 (ebook)
Subjects: LCSH: Internet of things--Security measures.
Classification: LCC TK5105.8857 .I65 2021 (print) | LCC TK5105.8857
(ebook) | DDC 005.8--dc23
LC record available at https://lccn.loc.gov/2020032513
LC ebook record available at https://lccn.loc.gov/2020032514

ISBN: 978-0-367-89387-3 (hbk)
ISBN: 978-1-003-03244-1 (ebk)

Typeset in Times LT Std
by KnowledgeWorks Global Ltd.

Contents

Preface

Internet of Things (IoT) is empowered by various technologies used to detect, gather, store, act, process, transmit, oversee, and examine information. The combination of emergent technologies for information processing and distributed security, such as cloud computing, artificial intelligence (AI), and blockchain, brings new challenges in addressing distributed security methods that form the foundation of improved and eventually entirely new products and services. As systems interact with each other, it is essential to have an agreed interoperability standard that is safe and valid. This book aims at providing an introduction by illustrating state-of-the-art security challenges and threats in IoT and the latest developments in IoT with cloud, AI, and blockchain security challenges. This book provides a comprehensive guide to researchers and students to design IoT-integrated AI, cloud, and blockchain projects. Various application case studies from domains such as science, engineering, and healthcare are introduced along with their architecture and how they leverage various technologies in cloud, AI, and blockchain. Moreover, more advanced topics are presented, and it is interesting for researchers in the field of IoT who want an overview of the next generation's challenges in IoT-integrated technologies that may occur in the coming years.

Chapter 1 provides an introduction to IoT architecture and relevant security protocols. The chapter also discusses integration management and security protocols. Apart from this, the impending challenges in IoT security are also outlined with the latest developments leading to plausible insights for solutions.

Chapter 2 covers the essential aspects of security challenges and threats in IoT. Furthermore, the chapter discusses the three-layer architecture involving hardware communication protocols and a layer of services and applications, security, integrity, and hardware issues, challenges, threats, risks, and solutions involving the deployment of efficient privacy, control, authentication, and integration methods to confront various series of malicious cyberattacks.

Chapter 3 discusses the role of cryptography and its challenges in integrating secured IoT products. This chapter also discusses the main challenges of cryptography and future directions of cryptography in integrating secured IoT products.

Chapter 4 explains the convergence of significant paradigms like blockchain and IoT that includes security and interconnectivity to modernize things. Moreover, it discusses the challenges of implementing blockchain and its related security issues in IoT.

Chapter 5 describes how artificial intelligence and machine learning (AI/ML) technology is being used to discover, manage, monitor, and protect all devices on a network. Various ML methodologies are discussed, including centroid-based clustering, hierarchical clustering, and classification techniques such as random forest.

Chapter 6 identifies some of these specific issues and discusses various approaches for mitigating security concerns. A classification system based on the Common Vulnerability Scoring System (CVSS) is presented. The technology applied is wide-ranging, including the integration of multi-vendor systems, AI/ML and the use of multiple database resources such as NVD/CVE and FDA.

Chapter 7 focuses on different security issues, requirements, and solutions for IoT-based WBAN healthcare systems with a discussion of various security metrics.

Chapter 8 introduces the blockchain concepts and their working principles. It covers in detail the notions of public ledgers, immutability, and consensus algorithms existing in a blockchain. It also analyzes the security flaws in IoT, which need to be addressed to secure them for crucial applications. This chapter also focuses on various security limitations of IoT and the usage of blockchain to overcome those issues and make them favorable for real-time applications.

Chapter 9 presents the concept of the industrial IoT along with its security challenges and applications. It covers the existing security issues in the area of cyber-physical systems and the evolution of industrial IoT with a secure design pattern.

Chapter 10 discusses the challenging factors in the cloud IoT integration model.

In Chapter 11, a homomorphic encryption scheme is described to authenticate the data from IoT devices to the cloud securely. This algorithm acts as privacy-preserving outsourced storage and computation in the cloud IoT model.

Chapter 12 describes the system to monitor the in-out activities of visitors to maintain the security of the laboratory and solve the occupancy detection problem.

The practical implementation of MQTT in IoT applications is discussed in Chapter 13. It is used to reduce transport overhead, protocol exchanges with minimized information, and network traffic, and it includes a mechanism to notify interested parties when an abnormal disconnection occurs.

Editors

S. Velliangiri earned his bachelor's in computer science and engineering from Anna University, Chennai; master's in computer science and engineering from Karpagam University, Coimbatore; and Doctor of Philosophy in information and communication engineering from Anna University, Chennai. Currently, he works as an associate professor at CMR Institute of Technology, Hyderabad, Telangana. He was a member of the Institute of Electrical and Electronics Engineers (IEEE) and the International Association of Engineers (IAENG). He specializes in network security and optimization techniques. He has published articles in more than 30 international journals and presented papers in more than 10 international conferences. He serves as a technical program committee and conference chair in many international conferences. He also serves as an area editor in *EAI Endorsed Journal of Energy Web*. He was the reviewer of *IEEE Transactions*, and Elsevier, Springer, Inderscience, and reputed Scopus indexed journals.

Sathish A. P. Kumar is currently an associate professor in the Department of Electrical Engineering and Computer Science at Cleveland State University (CSU). His research interests are in cybersecurity, machine learning, distributed systems, and related applications. Prof. Kumar has served on the program committees of numerous international conferences and as a reviewer for *IEEE Transactions*, including *IEEE Transactions on Mobile Computing, IEEE Transactions on Cloud Computing*, and *IEEE Transactions on Services Computing*. Prof. Kumar has published more than 50 technical research papers in journals and international conference proceedings.

P. Karthikeyan obtained his bachelor of engineering (BE) in computer science and engineering from Anna University, Chennai, Tamil Nadu, India in 2005 and received his master of engineering (ME) in computer science and engineering from Anna University, Coimbatore, Tamil Nadu, India in 2009. He earned his PhD from the Anna University, Chennai, in 2018. He is skilled in developing projects and carrying out research in the area of cloud computing and data science with programming skills in Java, Python, R, and C. He has published articles in more than 20 international journals with good impact factor and presented papers in more than 10 international conferences. He was the reviewer of Elsevier, Springer, Inderscience, and reputed Scopus indexed journals.

Contributors

L. Sherly Puspha Annabel
St. Joseph's College of Engineering
Chennai, Tamil Nadu, India

P. Avirajamanjula
Department of EEE
Prist University
Thanjavur, Tamil Nadu, India

R. Biswas
Applied Optics and Photonics
    Laboratory
Department of Physics
Tezpur University
Assam, India

D. Citharthan
Christ the King Engineering College
Coimbatore, Tamil Nadu, India

N. Dhanasekar
Department of EEE
A.V.C. College of Engineering
Mayiladuthurai, Tamil Nadu, India

Sudhakar Hallur
KLS Gogte Institute of Technology,
    VTU
Belagavi, Karnataka, India

Yogesh M. Iggalore
METI M2M India Private Limited
Mysore, Karnataka, India

S. Rakoth Kandan
Department of CSE
Jayamukhi Institute of Technological
    Sciences
Warangal, Telangana, India

T. Karthikeyan
Annamacharya Institute of Technology
    and Sciences
Andhra Pradesh, India

N. V. Kousik
Galgotias University
Uttar Pradesh, India

Roopa Kulkarni
DSATM, VTU
Bengaluru, Karnataka, India

S. Suresh Kumar
QIS College of Engineering and
    Technology
Ongole, Andhra Pradesh, India

N. Mahadevan
Department of ECE
Coimbatore Institute Technology
Coimbatore, Tamil Nadu, India

Gnanaprakasam Pandian
Ordr.net
Santa Clara, California

Saswati Paramita
Dayananda Sagar University
Bengaluru, Karnataka, India

Prashant Patavardhan
Dayanand Sagar University
Bengaluru, Karnataka, India

M. Poongothai
Department of ECE
Coimbatore Institute Technology
Coimbatore, Tamil Nadu, India

J. Premalatha
Kongu Engineering College
Erode, Perundurai, Tamil Nadu, India

R. Arshath Raja
ICT Academy
Tamil Nadu, India

D. Palanivel Rajan
CMR Engineering College
Hyderabad, Telangana, India

Vani Rajasekar
Kongu Engineering College
Erode, Perundurai, Tamil Nadu, India

K. Sathya
Kongu Engineering College
Erode, Perundurai, Tamil Nadu, India

Kripa Sekaran
St. Joseph's College of Engineering
Chennai, Tamil Nadu, India

Mark Sue
Ordr.net
Santa Clara, California

M. Varatharaj
Kathir College of Engineering
Coimbatore, Tamil Nadu, India

Vivek Vinayagam
Ordr.net
Santa Clara, California

Brian Xu
Ordr.net
Santa Clara, California

N. Yuvaraj
ICT Academy
Tamil Nadu, India

1 A Brief Overview of IoT Architecture and Relevant Security

R. Biswas
Applied Optics and Photonics Laboratory
Tezpur University
Sonitpur, Assam, India

CONTENTS

1.1 INTRODUCTION

Of late, the world has seen a large upsurge in the number of smart devices. With the advent of industrial growth as well as other relevant fields, the number of devices has been rising enormously. It is expected that the number of connected devices will surpass 150 billion by 2025. The enormity cannot be solely attributed to growth in human population. Rather, if we dig deeper, there is another reason for this. The daily utility devices and the functional technologies rooted directly on factory floor are interconnected gradually across the entire globe. This leads to an assembly of interconnected things where there is human-to-machine (H2M) interaction and also machine-to-machine (M2M) interaction. All these now come under one big umbrella – Internet of Things (IoT). Accordingly, we can define IoT as 'An omnipresent network capable of monitoring as well as controlling physical parameters through collection, processing and eventual analysis of data being produced by sensors or smart objects'. This chapter gives a general overview of the IoT architectural growth and also attempts to highlight the allied protocols along with security measures. Apart from this, this chapter provides a formidable framework for managing IoT security [1–8].

Since its emergence, IoT has been spreading fast. Nowadays, most of the enterprises are IoT-enabled, including healthcare, automobile industry, heavy machineries industry, and so on. Figure 1.1 shows the stages that IoT has been through and will be assuming in near future. The nodes represent the active smart sensors such as mobile phone, tablets, laptops, smart houses, etc., as illustrated in Figure 1.1.

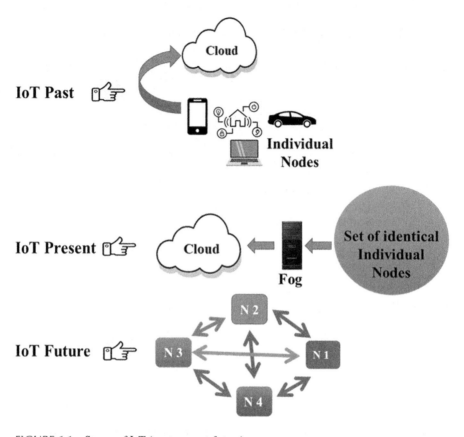

FIGURE 1.1   Stages of IoT (past-present-future).

Interestingly, the number of things connected to internet exceeded human population in 2008. Thanks to the tremendous speed of adoption rate of IoT, IoT is growing at a larger scale. Again, the shift from consumer-based Internet Protocol version 4 (IPv4) internet for nodes to operational technology-based IPv6 internet of M2M interaction, which is inclusive of smart sensors, clustered system, etc., has been recently driving its scalable growth.

Although the concept of IoT and M2M communications is sometimes mistaken to be the same, however, M2M precisely remains as a subset of IoT [3–4]. Thus, IoT in its bigger umbrella encompasses the following:

a. Radio-frequency identification (RFID)
b. Machine to human communication
c. Location-based services (LBS)
d. Lab-on-a-Chip (LOC)
e. Augmented reality
f. Robotics
g. Automobile telematics

Among the aforementioned technologies, majority are the outcomes of developments in military and allied supply chains with a common goal to connect embedded sensors through a thread of communication intelligence. The data is run over a mix of wired and wireless networks. Precisely, the whole architecture is embedded through smart objects in the form of intelligent entities such as automobiles, mobile phones, laptops, thermistors, etc. IoT comprises a very vital component in its architecture, known as *embedded and distributed intelligence*, which takes care of several key aspects including data collection, network resource preservation, and closed loop functioning. Known also as 'fog computing', *distributed intelligence capability* prioritizes proximity to source rather than to a centralized system. As an extension of the cloud paradigm, it integrates cloud into the physical world of things [7–9].

1.2 IoT PROTOCOLS

Over the ages, IoT has been adhering to several protocols developed since its inception. The protocols scale down to all physical layers in conformity with the International Organization for Standardization (ISO) with an objective to facilitate the functioning IoT devices with ease. For instance, we can refer to messaging protocols comprising the Constrained Application Protocol (CoAP). Sometimes, there is implementation of highly extensible routing protocols. Depending on the demand, it may troubleshoot Lossy Networks as well as Low-power. In such cases, IoT exhorts to Routing Protocols [11–15]. The main criterion behind these developments of protocols is to preserve energy, which is further accompanied by minimal computing and memory requirements. Meanwhile, IPv4 has been one of the dominant protocols from the beginning. However, there is a limitation of connecting billions of devices to it. As a result, it is then taken over by IPv6 that has the capability of handling a large number of IoT devices. As a result, assorted security considerations and implications in connection to IPv6 fundamentally build the security of IoT [1–4]. The Protocol Tree of IoT is provided in Figure 1.2.

1.3 IoT INTEGRATION

In the preceding two sections, we discussed the growth of IoT as well as different protocols applied in case of IoT. As mentioned in Section 1.2, there exist billions of IoT devices which remain connected. With the growing number of interconnected smart devices as well as the allied peripherals, there are pretty chances of mismanagement or synchronization within the nodes. In these cases, IoT integration comes as a saviour for alienating such type of problems. The definition of IoT integration can be expressed as 'A synergistic amalgamation of novel IoT devices, IoT data, IoT platforms and IoT applications with IT assets towards achieving end-to-end IoT business solutions' [7–8].

Apparently, IoT endures tremendous trust as well as concern for the consumers. So, many enterprises are reliant on it. As a result, the complexities incurred by IoT integration to conventional IT infrastructure are often underestimated. As for beginners, IoT yields enormous volume of data that the enterprises have to deal with.

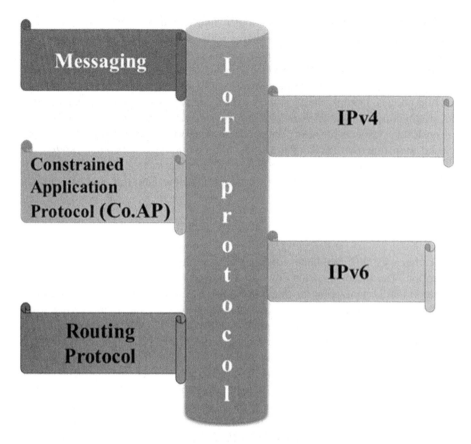

FIGURE 1.2   Protocols obeyed by IoT. (Goal: Preservation of energy.)

Accordingly, there occurs an exponential inflation in the IoT budget, which is related to covering the expenses entailing integration risks. As such, lot of enterprises initiate working without apprehending the challenges associated with IoT integration. Another noteworthy point is the obscurity lying in the allied protocols, which makes it difficult to synergize. Let's have a look at the crucial points that make integration with IoT challenging. Figure 1.3 illustrates these challenges.

From Figure 1.3 it is evident that certain features of IoT make its integration quite challenging. For instance, there is a humongous number of connected devices which are participating in concomitant exchange of data, thus resulting in a huge number of end points. Again, this results in accruing and doling out a stream of data emanating from a huge number of smart sensors. Moreover, some enterprises rely on traditional integrations based on batched data, which come as a hindrance in the end. Dynamic end points on the other hand possess a tendency of expansion, thereby making the job harder for the conventional integration to manage a fixed number of end points. Again, the nodes, i.e., the smart sensors in IoT, sometimes utilize newer communication protocols. Consequently, this induces pressure on

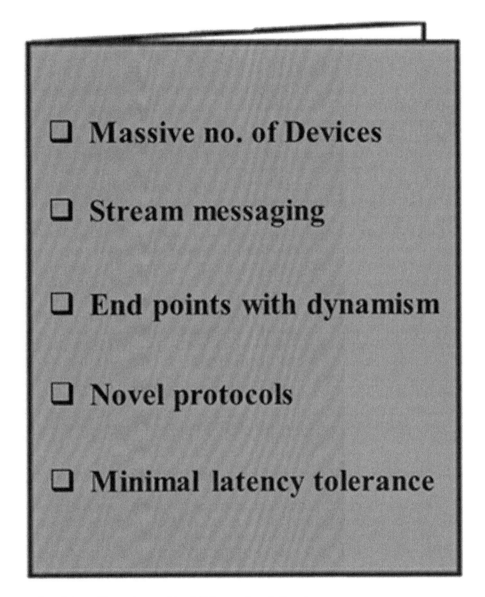

FIGURE 1.3 Key points making IoT integration challenging.

enterprises, opting for integration platforms with the capability of tuning seamlessly with novel IoT protocols [7–8]. Above all, there should not be any latency tolerance. Accordingly, they should come up with potential of handling superfast processing of data.

When we want to attain credibility as well as full control with our IoT projects, it is necessary that we should take care of it from the very beginning. This necessitates a one-to-one correspondence with data practices as well as business processes of the enterprise under consideration. Again, with higher scaling of IoT capabilities, the task of integration can become complicated. This can be tackled with cloud-based platform and can help in managing multilevel workflows, IoT projects with shared components, general purpose translation, etc. For instance, two or more data sources being hosted on-premises can be handled well provided there is in-house hosting of middleware technology. This, in turn, succours the small- and mid-sized farms that have invested heavily in on-premise applications.

There is another avenue where IoT integration can be sorted out. Through adoption of effective communication approach, we can manage integration of IoT projects with ease. With a view to improving the core processes, there should be a planned execution of communication. This effective planned approach should take into account items such as quantity of assorted devices, device type, and knowledge about the range of commutation technologies as well as tolerance capacity. It should be further strengthened by device autonomy though localized computing. As mentioned previously, the massive data as well as enormous dynamic end points make the security aspect of IoT more thought provoking. Accordingly, end-point encryption becomes a mandate now. However, different format encryption again raises the security concern for IoT. Even though all these are taken care of, still the job is not done. Suppose we are running our enterprise's analytic engine to its fullest potential, however, we are quite ignorant of it and we feed our first IoT project. In such cases, we have to take additional measures to provide space to the huge data that IoT devices will be generating. Accordingly, when feeding an existing analytics system with data streams, we should have a mechanism to address the challenges coming with data format as well as mapping. This requires a fully optimized and equipped IT sector.

These integration of challenges can be handled with the following procedures [8–16]:

a. **Implementation of Application Programme Interface (API) approach**
   Since IoT projects are mostly dependent on mobile and cloud computing, the adoption of API is quite relevant as most computing-based technologies are already API centric. Nevertheless, we should not mistake this to be solely API approach. Again, here is cautionary point, as API alone is not able to bear all capabilities enticing security and scaling of integration occurring in large distributed systems.

b. **Proper assessment of communication for IoT devices**
   When we start up our IoT project, it is necessary that we should build up an effective communication approach. For instance, our first job is to find out the way of communication as well as the connected technology. The technology may be cellular networks or short-range wireless. At the same time, it is imperative to look upon the number and variety of the connected nodes. Once it is done, we should then move over to network topology that suits our best requirement for device autonomy involving either edge computing or gateways. When we take care of these aspects, we can then decide the suitability of our approach to a bundled IoT platform or for network-assorted IoT.

c. **Enabling cloud towards data and process integration**
Through this step, we advocate integration of IoT platforms with core business processes. Sometimes, the built-in integration capabilities are adequate for initial level. However, scalability induces shortage of capabilities. In such levels, it is justifiable to resort to cloud-based approach, such as through adoption of Integration Platform as a Service (iPaaS) platform. This, in turn, helps in complex integration as well as accommodating more complex IoT resources.

d. **Selective use of traditional software**
If we have an IoT project which is basically rooted in on-premises, we can judiciously use conventional on-premises middleware. Although they are not adaptable to cloud services integration, they can suffice the on-premises IoT project.

e. **Use API management tools**
When we use APIs, then there is every possibility of their proliferation. In such circumstances, it is advised that there should be a provision of addition of API management tools/solution. The general opinion is that the abilities of managing API broadly vary corresponding to either several IoT platforms or middleware. As such, management tools of API ensure security as well as dependable scalability. They also smoothen diverse requirement for IoT projects.

1.4   IoT SECURITY

In the previous section, we discussed challenges associated with IoT integration as well as the plausible measures. However, smoothened integration does not imply total security in IoT projects. Due to the colossal number of smart objects in IoT as well as voluminous data streaming, each sector of IoT can be under total threat if the security issues are not dealt with caution. On several occasions, challenges arise due to disruption of traditional model. Let's have a look at the challenges associated with IoT security:

i. Absence of physical security with normally petite and inexpensive devices.
ii. Lack of advanced computing platforms since the existing may be constrained in memory as well as computing resources which are basically ascribed to limited security computing abilities, inadaptability of encryption algorithms to higher processing power, as well as low CPU cycles vs versus effective encryption.

In some cases, there is precedence of autonomous operation in field without any provision of backup when there is a chance of losing the primary connection. Again, some IoT projects are implemented without taking into account the network availability, which thus lengthens the on-boarding time. Even though it is operational, it must be remotely manageable during or after boarding. Meanwhile, there should be scalability and management of billions of entities in the IoT system. Figure 1.4 highlights the measures that can be taken to solve security issues with IoT.

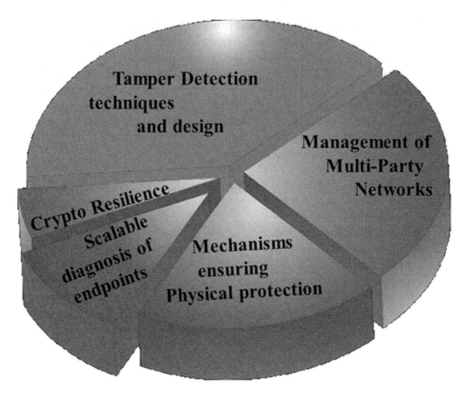

FIGURE 1.4   Plausible measures in case of IoT security.

As shown in the figure, it is imperative that there should be scalable diagnosis of billons of entities as well as end points and their effective management. For example, it may start from an individual to group level. In case of individuals, we may cite about smart meter or thermistor. In case of groups, we can mention bulbs used in the lighting of a house. Now, the scalability challenges arise as a tussle between individual and group entities. In such cases, the location-specific end points can offer best solution instead of individual identifier.

Again, when we have to deal with an IoT project catering to varied consumers, it can indulge in multiparty networks. In such cases, the management should be done in such a manner where there is credibility of provisioning resource for the consumers. Common nodes with varied parties with different *modus operandi* should also be handled with stringent conditions. Another crucial measure that can be adopted is crypto resilience. Nevertheless, the design of a tamper detection is the need of the hour. Again, when all these measures are taken, the approaches will be futile as there has been no physical protection. The chances of theft as well as intentional movements of nodes should be kept minimal in order to ensure a reliable IoT system.

It has now become quite clear that there is no option of single use or single ownership of the engaged IoT entities. The nodes and working protocols sharing and consuming belong to varied ownership, policy, managerial, and connectivity domains.

This demands concurrent, equal sharing and access to a number of data consumers and controllers. In such circumstances, the retention of privacy and exclusivity of data is of topmost priority. If we have a limited resource IoT, then it must adhere to some norms as a security measure including:

a. Secure authentication to multiple users
b. Availability of data to multiple collectors
c. Synergistic management between data access
d. Survival adaptability to unknown hazard

Now, let's discuss a real problem in order to understand the paramount importance of data access. Suppose, in an IoT-enabled crucial industrial process, there is exclusive dependence on precise measurement of temperature. In such crucial point, in case there arises a denial-of-service (DoS) attack for that particular end point, there ought to be an alert mechanism for the process collection agent. At this juncture, the IoT system should be capable of sourcing data from a standby loop or deferral of the transmission system. However, one noteworthy point is that the IoT system must be equipped so that proper identification can be accomplished between DoS attack driven data loss and loss of the connected device in order to avoid probable catastrophe. It must also be able to distinguish between loss of data due to an on-going DoS attack and loss of the device due to a catastrophic event in the plant. Here, we can mention enabling of IoT with machine learning.

In order to tackle the greatly diverse IoT environment and the related security challenges, we can plausibly implement a four-layer framework. The first two layers come with two As – they incorporate 'Authentication' and then 'Authorization'. These two As come under strict surveillance of 'Network-Driven Policy' which can be referred to as the third layer. The top layer comes as a big umbrella known as 'Secure Analytics', which is solely responsible for visibility and control.

1.5   IoT CHALLENGES AHEAD

Although IoT has come a long way, there is no denial of the fact that the threats as well as challenges are also piling up with the advent of technology. One major problem faced by IoT is lack of interoperability and unclear value propositions. Even after wide use, traders as well as consumers are afraid of heavy use of IoT due to some bottlenecks such as misappropriation of value propositions. Many opine that it lacks interpolability, while others question about the extent of adoption. Is it too early?

As per reliability measures of IoT, it should adhere to the following fundamental objectives:

a. Data confidentiality: Assurance of non-use by third parties.
b. Data integrity: Prohibition of intentional/unintentional corruption of stored and transmitted data.
c. Non-repudiation.
d. Data availability: Ease of access of stored and transmitted data.

Although being lucrative at first glance, there are many things which hinder its reach, including inherent uncertainties and lack of historical precedence. While some fear about dearth of feature of true capturing of IoT, others raise doubts about the organizational capabilities. The market dynamics play a pivotal role for use of IoT as many potential clients are waiting for further growth so that it can be deployable.

In summary, this chapter gave a brief overview of IoT in the context of integration and security measures. Starting from the growth of IoT, we elucidated the key points which are ruling the current IoT-enabled systems. The protocols in IoT were also briefly highlighted. Afterwards, we moved over to integration of IoT. Stating its considerable importance, we outlined the measures which can lead to effective management of synergistic integration, followed by the description of security issues that pervade the IoT sector. We also briefly discussed a plausible framework to combat security concerns related to IoT-enabling four-layer adaptation. While we cannot rule out the vastness associated with the security implications for IoT/M2M build, however, it can be a viable measure of decomposing an IoT/M2M security framework. The IoT industry is at its full bloom and continuing its large expansion and hence security threats cannot be denied. As a result, there is a need for new security protocols and diagnosis measurements which can lead to enhanced adaptability. Machine learning and artificial intelligence may act as a boon for combating key issues plaguing IoT security.

REFERENCES

1. Xiao, L., Q. Yan, W. Lou, G. Chen, and Y. T. Hou, "Proximity-based security techniques for mobile users in wireless networks," in *IEEE Transactions on Information Forensics and Security*, 8(12) (2013): 2089–2100.
2. Xiaohui, X., "Study on security problems and key technologies of the internet of things," in Proceedings of IEEE Fifth International Conference Computational and Information Sciences (ICCIS), (2013) Hubei, China.
3. Zhou, J., Z. Cao, X. Dong, and A. V. Vasilakos, "Security and privacy for cloud-based IoT: challenges," *IEEE Communications Magazine*, 55(1) (2017): 26–33.
4. Aleisa, N., and K. Renaud, "Privacy of the internet of things: asystematic literature review," in Proceedings of 50th Hawaii International Conference on System Sciences, (2017), Waikoloa, HI.
5. Kanuparthi, A., K. Ramesh, and S. Addepalli, "Hardware and embedded security in the context of internet of things," in Proceedings of the 2013 ACM Workshop on Security, Privacy & Dependability for Cyber Vehicles, (2013), 61–64, Berlin, Germany.
6. Tan, Z., A. Jamdagni, X. He, P. Nanda, and R. P. Liu, "A system for denial-of-service attack detection based on multivariate correlation analysis," in *IEEE Transactions on Parallel and Distributed Systems*, 25(2) (2013): 447–456.
7. http://techgenix.com [As accessed on 23//03/20].
8. https://www.gartner.com [As accessed on 23/03/20].
9. Dubey, H., A. Monteiro, N. Constant, M. Abtahi, D. Borthakur, L. Mahler, Y. Sun, Q. Yang, U. Akbar, and K. Mankodiya, "Fog computing in medical internet-of-things: architecture, implementation, and applications," in *Handbook of Large-Scale Distributed Computing in Smart Healthcare*, (2017). Springer: 281–321.

10. Gia, T.N., M. Jiang, A.-M. Rahmani, T. Westerlund, P. Liljeberg, and H. Tenhunen, "Fog computing in healthcare internet of things: a case study on ECG feature extraction," in *2015* IEEE International Conference on Computer and Information Technology, (2015): 356–363.

11. Bandyopadhyay, S., M. Sengupta, S. Maiti, and S. Dutta, "A survey of middleware for internet of things," in *Recent Trends in Wireless and Mobile Networks*, (2011). Springer: 288–296.

12. S. N. Swamy, D. Jadhav, and N. Kulkarni, "Security threats in the application layer in IoT applications," in *2017* International Conference on I-SMAC (IoT in Social, Mobile, Analytics and Cloud)(I-SMAC), (2017): 477–480.

13. Mosenia, A., and N. K. Jha, "A comprehensive study of security of internet-of-things," in *IEEE Transactions on Emerging Topics in Computing*, 5(4) (2017): 586–602.

14. Yu, W., F. Liang, X. He, W. G. Hatcher, C. Lu, J. Lin, and X. Yang, "A survey on the edge computing for the internet of things," in *IEEE Access*, 6 (2018): 6900–6919.

15. Lin, J., W. Yu, N. Zhang, X. Yang, H. Zhang, and W. Zhao, "A survey on Internet of Things: Architecture, enabling technologies, security and privacy, and application, in *IEEE Internet of Things*, 4(5) (2017): 1125–1142.

16. Karthikeyyan, P., and S. Velliangiri, "Review of blockchain based IoT application and its security issues," in *2019 2nd International Conference on Intelligent Computing, Instrumentation and Control Technologies (ICICICT)*, (2019): pp. 6–11.

2 Introduction, Security Challenges, and Threats in IoT

Sudhakar Hallur
KLS Gogte Institute of Technology, VTU
Belagavi, Karnataka, India

Roopa Kulkarni
DSATM, VTU
Bengaluru, Karnataka, India

Prashant Patavardhan
Dayanand Sagar University
Bengaluru, India

CONTENTS

2.1   INTRODUCTION

Applications involving wireless communications involve highest security risks and threats. Majority of the online applications now go wireless involving Internet of Things (IoT) as a technology to communicate to their respective destination. The IoT technology is envisioned to grow rapidly, providing strong support, guaranteed, and easily accessible services in the imminent years because of the expansion of the communication technology, the availability and utilization of the services, devices, computing systems, and applications.[1]

2.1.1   THREE-LEVEL ARCHITECTURE

IoT architecture is a three-layer system providing services and applications. A short depiction of each layer is given underneath. Figure 2.1 depicts the three-level IoT architecture.[3]

IoT Application and Security Layer	Logistics Management and Security	Intelligent Traffic Management and Security	Remote Device Support and Security	Environment Monitoring	Smart Devices Lookout and Security	Various other application security	**Application Support and Security Layer**
Application Support Layer	Intermediate Support Platform	Data and Cloud Computing Support	Service Support Platform and Management	Data Development and Security Platform	Other Application Support and Security Platform		
Core Network Attributes	Internet Source Security and Management	GPRS and EDGE Security	3G / LTE / 5G Network Security	High Speed Network Security	Other Network Security and Management		**Network Layer**
LAN and Network Access and Security	Wi - Fi Network Management and Security	Bluetooth Security	VPN Security	Ad hoc security	LAN and WLAN Security		
Perception / Conscious Network	Wireless Sensor Network Management and Security	MEMS and NEMS Management and Security	Radio Frequency Identification Management and Security	Global Positioning Sensor Management and Security	RADAR and Waveform Management	Numerous other Sensor Management	**Perception Layer**
Perception / Conscious Nodes							

FIGURE 2.1   The three-layer IoT architecture.

2.1.1.1   Perception (Edge) Layer

This layer is responsible for the acquisition and conversion of the physical data, its properties, and characteristics using embedded intelligent nanochip sensors.

2.1.1.2   Network Layer

This layer processes huge quantities of the data received from the perception layer and transmit it to the application layer utilizing remote innovations technologies and stores it onto a reliable huge cloud data storage having dynamic processing capabilities.

2.1.1.3   Application Layer

This layer is equipped with the tools needed by the developers for implementation and also constitutes the front end of the whole architecture.

2.1.2   IoT Protocols

Divergent application protocols like CoAP, DDS, XMPP, REST, AMQP, and MQTT have been proposed as a solution for IoT communication problems. An IoT protocol that guarantees a solution for fragmentation problems is the MQTT protocol that reduces the delay as compared to other protocols. For the applications that provide support to TCP/IP protocol suite, a set of IoT protocols such as Constrained Application Protocol (CoAP), Representational State Transfer (REST), Message Queue Telemetry Transport (MQTT), Extensible Messaging and Presence Protocol (XMPP), Data Distribution Service (DDS), MQTT for Sensor Networks (MQTT-SN), etc. Further, the communication patterns between the various entities comprising of IoT may be broadly categorized into three main classes: machine-to-machine (M2M), machine-to-server (M2S), server-to-server (S2S).[2]

2.1.3 IoT Elements

Wireless technologies such as RFID, ZigBee, etc. play a crucial role in providing end-to-end IoT communications by using radio frequencies.[19] The IoT such as RFID tags attached as stickers to objects may have inbuilt power (active) or may depend for power on some natural resource such as solar, etc. (passive) using which collaborate within themselves to create a network, which will be headed by a leader elected on different constraints. These sensors are called nodes and have huge applications. The main component of IoT is IoT middleware, decomposed into various simple components following a service-oriented approach permitting the use of any hardware as well as software. A short description of each designated layer in IoT is as mentioned below.

2.1.3.1 Application

The topmost layer providing a graphical user interface to make it easier for the user to interact with the nodes. It also provides a definite structure to the nodes that are to be interconnected and data to be extracted from.

2.1.3.2 Service Composition

The topmost layer in the middleware, whose primary function is to aggregate all individual available services to build the application as a whole.

2.1.3.3 Service Management

A middle layer of the middleware that manages all the components, services, and objects in the framework whose function is to provide advanced services such as setting up, monitoring, and managing the connections.

2.1.3.4 Abstraction

Any sensor connected to any object interacts in the binary language. The abstraction layer consists of all the devices with access to each one of them and also interprets various techniques and orders involved in the arrangement gadgets to collaborate with real-life objects. This layer ensures confidentiality, fidelity, and management, which assists in concealing data to ensure secure communication.

2.2 SECURITY RISKS AND THREATS

Nature-wise risks are classified into active and passive threats. The attacks are majorly classified into 3 categories layer-wise. Some of them are as mentioned below.

2.2.1 Physical Attacks

2.2.1.1 Tampering of Nodes

The adversary causes severe damage to the IoT sensor node by supplanting the piece of its equipment or the whole node itself or even examine or interrogate the nodes electronically to obtain access, so as to change the sensitive data, which may result in the defective operation of higher communication layers.[33]

2.2.1.2   Malicious Node and Code Injection

Here, a new pernicious node or a little of noxious code is infused amidst two or more nodes of the IoT framework, thus controlling their operations and limiting the flow of the data to and from the nodes and their respective destinations.[33]

2.2.1.3   Physical Damage

The attacker may physically harm, damage, and destroy the IoT elements for personal gain affecting its service availability. This sort of attack takes into control and affects the security of the zone that has the IoT framework.

2.2.1.4   False Data Injection

This attack generated via different techniques with restricted data of the power system topologies consists of some pernicious data infused or modified within the measurement meter data,[25] which may go unnoticed by the present state estimation strategies.[26]

2.2.1.5   Eavesdropping

The risk of data being intercepted by an attacker during generation or extraction of information by the IoT elements during communication without the sensor or communication medium having knowledge about this transfer is called eavesdropping. No data is harmed in this attack. This risk can be minimized by scrambling and encrypting the codes and the data.[19]

2.2.2   Network Attacks

2.2.2.1   Data Security

The difficulties can be pulled back using two approaches: data accumulation and data anonymization.[20] The usage of duplicate tags by the attacker may lead to easy access to the confidential data from the legitimate nodes. It can be resolved by shielding the tag, so that it cannot be duplicated. Blocking may be blocking a tag and causing a denial of service (DoS). This may be minimized by detecting it at an early stage.[19]

2.2.2.2   Replay Attacks

These attacks aim to fetch the authentication user or system information. In IoT, the replay attacks to intercept and capture the usage pattern and replay the similar data to perform an undetected intrusion.[21] Further, the attacker may inject wrong and irrelevant data into the system, leading it to over energy consumption and inaccurate prediction.[22]

2.2.2.3   Traffic Analysis Attacks

An attacker may track down the private data spilling out of the IoT nodes on account of their remote attributes.[19] Also, in any form of network attack, the adversary initially tries to capture some network information prior to the attack. These attacks are usually done by scanning the ports, packet sniffing, etc.

2.2.2.4 Sinkhole Attack

The adversary draws the complete traffic from IoT sensor nodes, making an allegorical sinkhole breaching the classification, confidentiality, and privacy of the data. It also denies access to the services to the network by dropping all the packets as opposed to sending them to the intended destination.[40]

2.2.2.5 Routing Information Attacks

A direct form of attack by the adversary that may use altering, spoofing, or replaying the routing information creates routing loops, complicating the network, sending false error messages, allowing or dropping traffic, partitioning the network, shortening or extending source routes, etc.[35]

2.2.2.6 Frequency Jamming Adversaries

Jamming is one of the common attacks done by mind-boggling ill-traffic used by the adversary to exploit or block the data, paralyze the operations, deny services of operations communication, and compromise the wireless environment.[19] Figure 2.2 depicts a jamming attack.

2.2.3 SOFTWARE ATTACKS

2.2.3.1 Phishing

The foe deals with the private information by parodying the confirmation qualifications of the client, for the most part through phishing sites or tainted messages.[3,33]

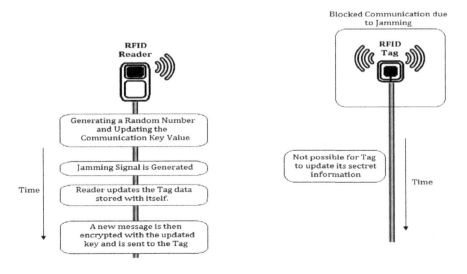

FIGURE 2.2   A jamming attack.

2.2.3.2   Blocking

Generation of a duplicate tag by an unauthorized person and blocking the intended tag causing DoS to the clients is known as Blocking. This may be detected at an early stage.

2.2.3.3   Worms, Virus, Spyware, Trojan Horse and Adware

An attacker infects the system with malicious software resulting in a variety of outcomes such as spying on the data, accessing the data via a backdoor entry, stealing the information, sending advertisements, such that clicking on it provides complete access to the user data, affecting other nodes, tampering data or even DoS.[33]

2.2.3.4   Malicious Software/Scripts

The client controlling the IoT gateway may be deceived by running executable active-x scripts or malware, which could result in data theft or a complete system shut down.[33] Malwares are categorized into three different types. They are pandemic malware, endemic malware, and contagion malware affecting the devices in the most intelligent way, in shortest time, and attaching itself to the legitimate hosts.[23,24]

2.2.3.5   Spoofing

Spoofing attacks are mainly classified into three types: IP spoofing, GPS spoofing, and ARP spoofing.[26] IP-based spoofing involves using a modified Internet Protocol address to surpass the security systems. GPS spoofing involves broadcasting inaccurate/spoof signals of higher signal strength than received from the source to delude the victims. Lastly, fake ARP data messages to connect the assailants MAC address with the IP address of the victim. Once all the spoofs are in place, all the information in the undermined framework will be under the complete authority of the interloper.[21]

2.2.3.6   Denial of Service Attacks

A Man-In-The-Middle (MIIT) attack makes use of ARP, which maps the address of the internet protocol to a MAC address of the victim. Once the control is gained over the network, the attacker acquires the original data, modifies it as per his requirements and then reverts it back to the destination.[21]

2.2.3.7   False Cloud and IoT

IoT involves the creation of fake cloud storage for storing and services to process the data. A cloud utility can either be a connection between an individual and an internet gadget or the last component to extricate the information from the sensor elements.[10] However, when the destruction or blockages are done by simply diverting the data from the original cloud to an alternative cloud, it is known as false cloud, where cloning is done to gain the control and access over the confidential data and then diverting it to the duplicated cloud to process it.[4,6] This false diversion is based on the filtering of the data packets based on Internet Protocol (IP) addresses and the destination ports. The packets appended with the cloud service address as their IP-addresses are replicated and forwarded to a false cloud server.[7]

2.2.3.8   Buffer Overflows and Reboot

Usage of fragmented packets in IPv6 communication and mapping it to a smaller frame size as per IEEE802.15x standard, reconstruction of the fragmented packets becomes necessary at the destination side. This leads to rebooting of devices, buffer overflows, and replay attacks on the destination side.[38,39]

2.2.3.9   Data Hiding

The attacker may exploit the data acquired by the nodes directly or may recover the data from the storage and hide it so as to showcase that the data is not available. This enables the attacker to control IoT devices. Through the data accumulation technique, the sort and proportion of data might be assembled.

2.3   SECURITY CHALLENGES

2.3.1   GENERAL IoT CHALLENGES

The IoT should consolidate different measures to affirm the identity of the individual entities requesting access to any data[3] and also should incorporate various security tasks and programs to ensure proper protection for the applications in the application layer.

A wide range of IoT systems combined with a high level of heterogeneity increases the security threats to device owners, machines, things, etc.[8] As the connected devices increase, the scalability problem also increases. However, in order to achieve an increased level of user satisfaction and acceptance, it is essential to provide increased security and confidentiality. Highly-secured authorization, rights to access, and authentication mechanisms need to be used to guarantee the integrity, trust, correctness, and confidentiality of the client's individual data.

Two important key characteristics in IoT infrastructures are self-healing and adaptation. They should be able to withstand any unforeseen situations that occur in the IoT environment. Thus, security issues and confidentiality have to be provided with the highest degree of flexibility to ensure the devices to have self-operational diagnostics, prevention of violations, and isolation.[9] Mechanisms have to be designed such that the data in an IoT environment has to be protected from an outside intruder but from inside threats. The more integrated systems and their back-end networks, the more the adversaries have an opportunity to munch over their malicious files on the network. Some of the general challenges include:

2.3.1.1   Access Control

IoT deals with streaming rather than processing digital data resulting in the problems of productivity and time constraints when dealing with traditional database management systems (DBMS).[8] Huge attention is given for receiving and storing data and providing various levels of security and confidentiality to the data received.[17] A hierarchical level of arrangement or security architecture of managing the resources and data takes into consideration the capacity of the storage device and limited processing power, where each user with only one key derives other necessary keys, reduces the storage costs, increases the confidentiality, integrity, and also the level of security for multiple nodes.[18] Continuous authentication on data streams (CADS) improves the scalability and performance. The service provider returns the results of the requests to the clients and also verifies the completeness and authenticity of the results obtained based on the information provided by the owner.

2.3.1.2 Insurance Concerns

Self-ruling gadgets include a lot of issues within themselves. But the data available can assemble them in a simpler way to survey dangers and also provides an opportunity for brand spanking new valuation models.

2.3.1.3 Lack of Common Standards

Various diverse IoT devices are interconnected, and each device needs a standard to be interfaced with the architecture. The lack of common standards among the IoT devices and protocols makes it difficult to achieve a broad industry adoption with one unified standard.[33]

2.3.1.4 Technical Concerns

The increasing data generation from the IoT devices represents a test to store, defend, process, and investigate it. The network ought to be prepared to withstand and manage high volume & density of the devices and should be capable of spotting and discriminating between permissible and scallywag devices.

2.3.1.5 Socio-Legal Concerns

There are no mechanisms to deal with these socio-legal considerations, such as which company or individual owns which device.

2.3.1.6 Need for Standardization and Interoperability

Most of the device manufacturers build their devices, making utilization of their underlying technologies that are unavailable to others. Thus, it is extremely necessary to standardize IoT interface parameters to render a higher degree of interworking between the objects.

2.3.1.7 Naming and Identity Management

IoT network is an extremely huge device/component connected network connecting billions and trillions of objects and devices, where each individual device or object needs to be addressed using their unique ID. Hence, an efficient and robust object naming mechanism is needed to progressively scan each object's identity. Hence, profiling and tracking each IoT Edge-connected device is a huge challenge. Also, assigning IPv6 address space to the objects support pervasive networking with increased reliability.

2.3.1.8 Confidentiality of Information and Secure Data Transmission

IoT allows seamless transfer of information from surveillance/capture devices to the cloud storage to enable live data analytics and investigation. IoT employs enormous identification technologies such as QR-codes, RFID Tags, etc. to identify the objects.

The safety of these tags needs to be ensured since they are embedded in the daily use of objects, carrying their data. Thus, it is essential to guarantee the safety, unauthorized access, and confidentiality of such data by using encryption algorithms and keys for generating illegible text. Despite of so much security, data security remains a challenge.

2.3.1.9   Network and Object Security

Since the data transferred is huge, congestion may occur in the network. Thus, the system used for transmission should be robust enough to manage the data arriving from numerous sensor nodes, to ensure no data is lost in the network. The network should ensure various efficient security measures such as encryption for the data to be transmitted, so as to prevent it from interference/corruption from any of the external sources.

2.3.1.10   Localization and Tracking

Localization may also be a threat as the IoT devices try to determine a person's location via space and time. Thus, designing and development of a protocol discouraging such activity become a major challenge.

2.3.1.11   Lack of Skills

Expertise and skills are very necessary in planning, design, development, implementation, and management of security of the data. The disturbance of any of these components may prompt the harm to the security framework in IoT. Also, it can cause the slow adoption of IoT technologies. The colossal test is the quantity of gifted individuals who can enough deal with IoT strategies, manage their difficulties is extremely constrained.[3]

2.3.1.12   Cost vs. Security Trade-Offs

In IoT, the requirement of special high-quality unit elements, hardware nodes and sensors, and software lead to a huge increase in cost. Using low-quality products lead to low security and increased potential risks.

2.3.1.13   Privacy Protection

IoT gives a provision of accessing the embedded devices from any location affecting the integrity and privacy of the sensitive data. Thus, some rules, regulations, and norms need to be followed to maintain a strategic distance from the infringement of security.[19]

2.3.1.14   Difficulty in Coping Up

The requirements of security and privacy for different devices are different. Also, as the devices and sensors increase, the interaction among them also increases. Thus, the difficulties of dealing with each of these focus in the system to boost the security additionally increase. Some of the security measures cannot be implemented due to a lack of storage capacity, power, and CPU.[3]

2.3.1.15   Restricted Infrastructure Resources

Since IoT gadgets ordinarily have low handling capacities and memory, it turns into a significant test for IoT programming engineers and equipment producers to plan the framework. Also, enough space must be provided for the security application to defend against security issues and threats. Also, as the sensor network size increases, the amount of data generated and stored also increases. Thus, data storage becomes a significant issue. Storing huge data peeks in the problem of providing security to the data. Also, it becomes difficult to back up the data when corrupted.[3]

2.3.1.16   Limited Resources on the Cloud

Sometimes, the data captured by the IoT sensors need to be backed up on the cloud for further need. As the data acquired from the sensors increases, the memory requirement for storage, CPU to process the data, algorithms to provide the security and confidentiality to data, scalability of the data, back-up mechanisms, virtualization mechanism, etc. on the cloud also increases. Increasing the resources increases the cost of implementation.

2.3.1.17   Weak Security Testing and Updating

As the IoT elements and sensors increases, testing all security-related aspects becomes difficult. Non-availability of the security patches for older devices and applying the latest patch of the update to new devices may not work properly. Thus, if the devices are not upgraded, they are prone to vulnerabilities such as hacking, etc.

2.3.1.18   Limited Power Sources and Node Capacity

Power is a crucial factor in IoT devices. The batteries do not easily get recharged, which results in an insufficient charge in the device resulting in failure of the network. Therefore, energy sources or batteries play a crucial role, especially in active sensor units.

2.3.2   LAYER SPECIFIC CHALLENGES

The IoT world view incorporates colossal parts and gadgets at various levels. Thus, a need to address security challenges at different layers arises. A scientific categorization of layer explicit IoT security challenges is as delineated in Figure 2.3. From Figure 2.3, the security challenges are categorized and distributed across three-layers of reference architecture.

2.3.2.1   Challenges in Application Layer

This layer is vulnerable to several attacks as described below.[33]

2.3.2.1.1   CoAP Security

The activity of Constrained Application Protocol (CoAP) is just like that to that of Hypertext Transfer Protocol (HTTP) when providing end-to-end security. A predefined structure is followed by the CoAP messages, as proposed in the standard RFC-7252.[34]

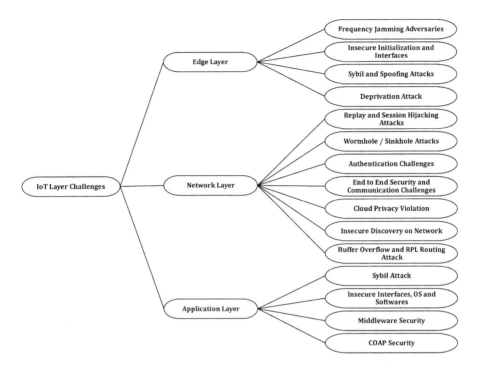

FIGURE 2.3 Layer specific challenges.

2.3.2.1.2 Shaky Interfaces

These interface issues are very common. IoT clients access their desirable services by means of web interfaces, which are focused as a weak point in their facilitating surroundings.[33]

2.3.2.1.3 Insecure Softwares and Operating Systems

Software or firmware embedded in IoT sensory-devices can have a vindictive application because of misconfiguration or attacks.[36]

2.3.2.1.4 Middleware Security

This aims at providing secure communication between various connected IoT components and interfaces. The middleware may interact with various environments increasing the security risks to the communication.[37]

2.3.2.2 Challenges in Network Layer

The significant functions of the privacy and security challenges of this layer include secured routing and communication and session management, applying different standards and involving various protocols of the network layer.[33] An attack taxonomy considering various network layer attacks is as listed below:

2.3.2.2.1 Insecure Nearest Node Discovery

Finding the close by working gadgets in a similar system of IoT condition utilizes Neighbor Discovery Protocol (NDP) in IPv6, which thusly finds the MAC locations of accessible routers, detects address duplication, and maintains the address resolution. Inappropriate authentication mechanism during the nearest node discovery may lead to a DDoS attack.[38]

2.3.2.2.2 Buffer Overflow

Space is needed to be reserved by the receiving nodes for incoming packets in the network layer. At times, an intruder may exploit this mechanism by sending inadequate or undesirable packets, leading to DoS attacks by discarding the useful packets since the limited buffer space being occupied by unwanted incomplete packets.

2.3.2.2.3 RPL Routing

This is one of the most open to attack protocol for low power and lossy networks.[31,39] In a Sinkhole attack, a noxious gadget demonstrates the ideal way to Base Station to alter the information extricated from other nodes, modify the packet payload, or even drop the message that may cause the delay.[40]

2.3.2.2.4 Sybil Attack

Some false nodes perform this attack by faking their identity. They even violate the client's information privacy, penetrating malware, spamming, or may make a phishing attack on neighboring surrounding nodes.

2.3.2.2.5 End to End Security

It provides the most secure way to guarantee a dependable correspondence between the two parties (transmitter and the receiver). The data privacy violation is the main security challenge to reach this goal.

2.3.2.2.6 Session Hijacking

It may lead to service denial to the users in the IoT environment by diverting the user data to some other domain. This occurs due to the fake messages being generated by the attackers to hijack the session.

2.3.2.2.7 Security Protocols

To provide better security services, differentiated and symmetric cryptosystems and key cryptosystems may be used, but they involve high computational overhead and delay. Thus, it is a challenge.[19]

2.3.2.3 Challenges in Edge Layer

The challenges in this layer occur due to limited direct access to maintain and configure the associated IoT elemental gadgets that incorporate hardware and data-link layer in the IoT environment leading to failure of the system functionality due to the threat propagation to further layers.[33]

2.3.2.3.1 Insecure Boot-Up

Every element in an IoT family is initialized during initial boot-up. The device is assumed to be more vulnerable to attacks if this procedure is not done properly. A malicious object may easily get access to the victim node settings and configuration, and tune it in an incorrect manner to exploit the data and other results. Thus, the devices in this layer need to be more secure for preventing unauthorized access.

2.3.2.3.2 Spoofing

It involves Sybil nodes impersonating themselves to degrade the functionality of the IoT environment. This may result in huge DOS for the legitimate nodes due to the consumption of network resources by the fake nodes. The challenge here is to detect false nodes and destroy them.

2.3.2.3.3 Insecure Interfaces

Insecure interfaces bring about the entire framework usefulness, including protective systems to be undermined by a malignant code in the system or the application layer.

2.3.2.3.4 Deprivation of Resources

Some attacks by the malicious nodes may result in keeping the IoT edge level devices to be awake all times and cause energy loss for the devices. At the same time, it may also restrict the legitimate nodes from fetching the resources.

2.4 INTEGRATION ISSUES AND CHALLENGES

The universe is connected via internet and thus a huge amount of data is generated. Huge storage space is needed on the cloud to save the data.[30] Also, huge processing and computational capabilities are needed to process the data acquired, which increases the cost.[30] Cloud computing with the incorporation of IOT devices and generating new patterns of data helps to achieve this. The integration and interfacing of IoT devices and elements with cloud is a tough task.[5] There are various challenges and difficulties that forestall the successful integration and performance of the cloud with IoT. Few of the challenges are:

2.4.1 PRIVACY AND SECURITY

One important issue in the transportation of data to be considered on priority is that the policies and authorization rules need to be followed by the authorized users to have access to the sensitive data. This becomes crucial when it comes to preserving the individual user's privacy, and particularly when data integrity needs to be guaranteed. The leakage of some sensitive information might also occur due to multitenancy. New challenges require special attention, such as the system is exposed to a number of new possible attacks.[32]

2.4.2   HETEROGENEITY

An important challenge in IoT is heterogeneity. This arises due to extensive variability among the IoT devices, their platforms and domains, and operating systems supported existing services, and standards that have to be followed. Cloud platforms and services accompany exclusive interfaces, taking into account resource integration dependent on explicit suppliers.[32] Moreover, the heterogeneity challenge is exacerbated when the end-clients receive the methodology of multi-cloud use, and in this manner, the services rely upon several providers to improve application strength and performance.

2.4.3   BIG DATA

The increasing data generated by the nodes leads to Big-Data processing, and thus, priority has to be given to the access, transportation, processing, and storage of it. It is only the cloud that facilitates the storage of the acquired information for an extensive stretch of time, by exposing it to complex handling and investigative techniques.[30] Managing and processing huge amounts of data produced is a serious issue, as the performance and outcome of the application are directly dependent on the specifications and features of the data management service. Developing an efficient algorithm to manage huge data is a tough task.[32]

2.4.4   PERFORMANCE

Transfer of enormous information produced from the IoT sensor elements to the cloud storage needs tremendous data transfer capacity and bandwidth. The important issue here is achieving the adequate performance of the network to transfer the obtained data to the cloud and that the broadband development isn't keeping pace with the storage and processing requirements on the cloud.[32]

2.4.5   MONITORING

Monitoring in cloud computing is mandatory with regards to capacity planning, data security, performance, management of resources, service level agreements (SLA), and troubleshooting. Additionally, some more challenges are influenced by the volume and velocity of the devices of the IoT.[30]

2.4.6 GIGANTIC SCALABILITY

The cloud-based IoT execution structures applications that are meant to analyze, investigate, and integrate the data originating from real-world via IoT devices. The huge size of the subsequent frameworks raises various new arduous issues. Accomplishing storage capacity requirements and computational ability has also become difficult.[32]

2.4.7 STANDARDIZATION

Lack of standards among various devices of IoT and cloud storage have created a critical issue in the integration of them.[32] Although many standardization approaches have been proposed for the deployment of IoT and cloud, it is the architectures, protocols, and APIs that need to permit the interconnections between diverse IoT gadgets new service generation, which in turn make up the cloud-based IoT paradigm. The mismatch among the norms presents the security challenges among IoT gadgets is a typical issue in IoT domain applications, especially in versatility and transportation conditions.[33]

2.4.8 ENERGY EFFICIENCY

The cloud-IoT applications that involve incessant data transfers from the IoT elements to cloud storage exhaust the energy of the node more quickly. So, efficient energy generation for processing of data and its transmission remains an open issue.[30] Usage of data caching techniques, reusing data, compression technologies, and efficient data transmission are some of the solutions to generate energy efficiently.[32]

2.4.9 SLA INTEGRATION ISSUES

Implementing common Service Level Agreements between the parties is a tougher job as the end-user requirements are different, and the services, priorities, guarantees, responsibilities, and warranties provided by the clouds may be different than of the user requirement. Thus, the integration of different services may not be convenient or feasible on the cloud.

2.4.10 NETWORKING ISSUES

Networking is a very important aspect in cloud integration as the cloud services completely rely on either wired/wireless networking to interact with a variety of devices. To support huge networking, shortest routing algorithms, optimized algorithms are required and the power required is huge. Also, the efficient usage of the energy of the network is required.

2.4.11 ASSIMILATION OF DEVICES AT RUNTIME

This is a problem that logically consolidates and configures the various available physical resources. Thus, numerous application platforms with distinguished characteristics are integrated with numerous IoT gadgets on the cloud and are adapted up to the adaptability of cloud just as restoration of devices.[30] Because of enormous adaptability, huge scalability and high maintenance, the endeavors in the prevailing environment fluctuate continually. Runtime osmosis is expected to store the data of device drivers related to the scalability of the device.[32]

2.4.12 INTEROPERABILITY AND NOMENCLATURES

The different streams of data are connected to different data sources, and the directing of data is done dynamically without any knowledge about the association of the entities. Additionally, the reconciliation of devices and terminologies provides the interoperability between the devices and trade the data. This exchanged data empowers the scope of abilities, for example, content-based routing, quality of service (QoS), and context-awareness. Consistent interoperability models are required to dissect the trouble of heterogeneous gadgets.[33]

2.4.13 MULTI-TENANCY

The association and the supply of the assets need a few ways to deal with the access and allocation of the resources among different clients.[33] Usage of shared resources appropriately facilitates the device cloud to cater the capabilities of access control, device tokens revocation, and dynamic granting. Every client acts as a resource supplier and furthermore as a client by retaining the devices. In this manner, clients substantiate against each other.

2.4.14 QUALITY OF SERVICE (QOS)

QoS of the network is measured based on the packet loss ratio, jitter, bandwidth, and delay.[27] The QoS reduces with huge data generation and inappropriate management of the resources. QoS also depends on the synchronization pattern of the data and its type based on its requirement.

2.4.15 STORAGE AREA

The area requirement, ability, and availability of the data storage matter a lot when it comes to storing delicate data like multimedia, i.e., photo, video, images, etc. These sorts of data need to be stored near to the area, such as it is easily accessible. Thus, accessing and processing time are reduced drastically.

2.4.16 RESOURCE ALLOCATION

Allocation of the resources poses problems with multiple IoT devices demanding for the resources and when deciding that particular resources required by an entity need to be fulfilled.[27] Based on the types and quantity of the sensors used, the frequency of generation of data need to be computed.

2.5 SOLUTIONS TO SECURITY AND PRIVACY ISSUES AND THREATS

2.5.1 GENERAL IoT SOLUTIONS

To speed up the process of the IoT, every security risks, threats, and challenges need to be provided with a solution to overcome with. Some of the solutions for challenges and threats are listed below:

2.5.1.1 Authentication and Confidentiality

It utilizes a client configurable embodiment component, such as Intelligent Service Security Application Protocol (ISSAP), that combines encryption with authentication, cross-platform communications, and signature to improvise the communication between the applications and the things.[8] The Datagram Transport Layer Security (DTLS) protocol is the first fully implemented two-way authentication scheme intended for IPv6 utilizing the 6LoWPANs standard and depends on the cryptographic algorithm RSA[10] which ensures data confidentiality, integrity, low packet delay, energy efficiency, and low memory load. Regarding integrity and confidentiality,[11] the key management system (KMS) protocols are appropriate for situations where computing power required is relatively low when compared with the utilization of public key cryptography.[12] A solution to the security attacks would incorporate an exchange model with signature encryption or an algorithm based on Boolean XOR operation plots that tend to address the security necessities of IoT.[13–16]

2.5.1.2 False Clouds

This false diversion of data can be prevented by initially checking by filtering the data packets IP address to a destination at the time of data extraction and sending it to destined port and then comparing it at the time of storing on the cloud. The data packets having the location of the legitimate cloud service same as their IP-address, are redirected to their respective cloud server.[7]

2.5.1.3 Device Assimilation

The problem of manual device assimilation can be resolved by automating the process and allowing it to manage the related process. The devices will likewise have the option to communicate with a solitary framework by means of a gateway-based methodology at the runtime utilizing automatic device assimilation.

2.5.1.4 Nomenclature and Interoperability

Common standards and algorithms for communication models, standards for design, and architecture need to be defined to interface the different IoT devices to the cloud. Also, this problem can be resolved by injecting them into the middleware of the device cloud at runtime.

2.5.1.5   For Multitenancy

The amount of client communication can be diminished by on-demand provisioning, characterizing the state models, and resolution-based metrics to permit access to the lowest level of architecture.

2.5.1.6   For Storage Area

The storage space may be increased, or a server for virtual storage might be furthermore allotted to the cloud to store multimedia data type.

2.5.1.7   Resource Allocation

The resource allocation problem can be overcome by delivering a sample packet from each newly added node, so as to indicate the node needs the resources to process or extract the data.

2.5.1.8   Centralized and Decentralized Architecture

Based on the attacker model applied to both centralized and decentralized IoT architecture models, the main challenges are studied, and appropriate solutions may be applied to deploy security mechanisms.[29] To overcome the issue of the data processing at only one place, decentralization of the cloud resources and processing CPU's may be done according to the user location, data characters, network density, etc. which can deal with the data that can be processed and stored independently. A centralized architecture may be a solution for the data in fields of defense, multimedia processing, etc., as the data is highly confidential and impossible to be distributed.[28]

2.5.2   LAYER-BASED SOLUTIONS

2.5.2.1   Application Layer

State-of-the-art solutions to security challenges involving applications based in any layer of IoT architecture are being presented in this section. Various solutions have been presented for the issues faced by the application layer of the IoT architecture.[33]

2.5.2.1.1   COAP Security

This type of attack usually affects the network layer and the application layer. Various encryption algorithms and security measures such as application proxy, DTLS, and resource directory are incorporated.

2.5.2.1.2   Insecure Interfaces

A major secure threat is via the user interaction with the user interface. Security issues such as password strengths and loose firewall may be answered by measures such as verification of the strength of the password, application gateway firewall installation, Secure coding such as SQLi injections, and cross-site scripting and also by using an encrypted channel by TLS.

2.5.2.1.3   Insecure Firmware

Two layers may be affected, i.e., network layer and application layer. Updating the firmware regularly via secure updates, usage of file signatures, encryption with authentication and validation, usage of lightweight encryption systems reduces the insecurity.

2.5.2.1.4   Insecure Middleware

There are three layers that might get affected via insecure middleware, i.e., network, transport, and application layers. The insecure middleware issue may be addressed by authenticating the communication channels, key management and distribution, various security policies, installation of various secure gateways, and M2M components.

2.5.2.2   Network Layer

State-of-the-art solutions to security challenges involving network-based in any layer of IoT architecture are being presented in this section. Various solutions have been presented for the issues faced by the network layer of the IoT architecture.[33]

2.5.2.2.1   Replay Attack

In IoT, the replay attacks intercept and capture the usage pattern and replay the similar data to perform an undetected intrusion.[21] Such attacks can be prevented by defining accurate authentication and validation parameters, and timestamps verification for packets, checksum verification using the hash value.

2.5.2.2.2   RPL Routing

This open attack for Low power networks can be addressed by monitoring the connected IoT devices and usage of lightweight encryption system.

2.5.2.2.3   Buffer Overflow

The buffer is overflowed by the waste and useless data, so no space is available for storage of the intended data. This condition can be prevented by the installation of intrusion detection system (IDS) modules.

2.5.2.2.4   Discovery of the Nearest Insecure Node

Inappropriate authentication mechanisms may lead to the nearest node discovery, which may also lead to a DDoS attack. This may be prevented by proper authentication and validation by encryption algorithms like elliptical curve cryptography (ECC) based signatures.

2.5.2.2.5   Sybil Attack

Discovery of the fake identities of the nodes can be accomplished by the analysis of user-interaction and graphical data and application of the access control list.

2.5.2.2.6   Sinkhole and Wormhole Attacks

Such attacks can be prevented by the usage of trust-level-based management system, by performing verification using hash systems, anomaly detection using intrusion detection system (IDS), device communication analysis, encrypted key-management, and signal strength monitoring.

2.5.2.2.7 End to End Security
The E2E security challenge may be addressed by the application of advanced encryption systems for authorization and authentication, for example, installing IPsec, network firewall, etc.

2.5.2.2.8 Session Hijacking
One of the most service denials is caused by service hijacking. Solutions to such attacks may be the usage of lightweight encryption systems and exchange key authentication after a specified time.

2.5.2.2.9 Guarded and Secure Communication
To ensure smooth, secure, and guarded communication asymmetric, and symmetric encryption methods, lightweight ticket granting system, packet payload dispatch encryption systems may be used.

2.5.2.3 Edge Layer
State-of-the-art solutions to security challenges involving edge based in any layer of IoT architecture are being presented in this section. Various solutions have been presented for the issues faced by the edge layer of the IoT architecture.[33]

2.5.2.3.1 Frequency Jamming
Frequency jamming challenges may be addressed by signal strength monitoring, computation of the packet delivery ratio, change of frequencies and locations, usage of error correction codes to encode packets, and various other methods.

2.5.2.3.2 Spoofing
Spoofing involves surpassing the security systems, broadcasting inaccurate higher strength signals, and usage of fake ARP data messages. This can be prevented by channel estimation and signal strength measurements.

2.5.2.3.3 Insecure Initialization/Configuration
Insecure initialization may be prevented by introducing artificial noise, configuring the data transfer rates between the nodes.

2.5.2.3.4 Shaky Interface
The shaky interface challenges may be tackled by avoiding the testing and debugging tools, usage of hardware-based TPM modules.

2.5.2.3.5 Sleep Deprivation
This challenge may be addressed by the usage of multilayer intrusion detection system (M-IDS) and artificial intelligence agents.

REFERENCES

1. X. Lu, Q. Li, Z. Qu, P. Hui, 2014. Privacy Information Security Classification Study in Internet of Things. In: *International Conference on Identification, Information and Knowledge in the Internet of Things.* pp. 162–165. doi: 10.1109/IIKI.2014.40.
2. A. Al-Fuqaha, A. Khreishah, M. Guizani, A. Rayes, M. Mohammadi, "Toward better horizontal integration among IoT services". In: *IEEE Communications Magazine*, vol. 53, no. 9, (September 2015): 72–79.
3. S. Singh., N. Singh, 2015. Internet of Things (IoT): Security challenges, business opportunities & reference architecture for E-commerce. In: *International Conference on Green Computing and Internet of Things (ICGCIoT).* pp. 1577–1581. doi: 10.1109/ICGCIoT.2015.7380718.
4. J. Stankovic. "A Vision of a smart city in the future." *Smart Cities*, vol. 1, no. 10, (Oct. 2013): 1–6.
5. A. Botta, W. de Donato, V. Persico, A. Pescape, 2014. On the Integration of Cloud Computing and Internet of Things. In: *International Conference on Future Internet of Things and Cloud. Proceedings.* pp. 23–30. doi: 10.1109/FiCloud.2014.14.
6. R. Kirichek, A. Koucheryavy, 2016. Internet of Things Laboratory Test Bed. In: *International Conference on Wireless Communication, Networking and Application. (WCNA).* LNEE. vol. 348. pp. 485–494, doi: 10.1007/978-81-322-2580-5_44.
7. A. Koucheryavy, Yim Chu-Hwan, L. Gilchenok, S. Moiseev, 2000. Overlay IPOP-network for Russia PSTN. In: *The 2nd International Conference Proceedings on Advanced Communication Technology, (ICACT-2000).*
8. S. G. Abdukhalilov, 2017. Problems of security networks internet things. In: *2017 International Conference on Information Science and Communications Technologies (ICISCT).* pp. 1–7. doi: 10.1109/ICISCT.2017.8188588
9. Y. Zhao, 2013. Research on data security technology in internet of things. In: *2nd International Conference on Mechatronics and Control Engineering (ICMCE).* pp. 1752–1755. https://doi.org/10.4028/www.scientific.net/AMM.433-435.1752
10. R. Roman, C. Alcaraz, J. Lopez, N. Sklavos. "Key management systems for sensor networks in the context of the internet of things". In: *Comput. Electric. Eng.*, vol. 37, no. 2, (2011): 147–159. https://doi.org/10.1016/j.compeleceng.2011.01.009
11. W. Du, J. Deng, Y. Han, P. Varshney, J. Katz, A. Khalili. "A pairwise key predistribution scheme for wireless sensor networks". In: *ACM Trans. Inf. Syst. Secure*, vol. 8, no. 2, (2005): 228–258. doi: M 1-58113-738-9/03/0010
12. Z.-Q. Wu, Y.-W. Zhou, J.-F. Ma. "A security transmission model for internet of things". In: *Jisuanji Xuebao/Chin. J. Comput.*, vol. 34, no. 8, (2011): 1351–1364. doi: 10.3724/SP.J.1016.2011.01351
13. Sabrina Sicari, Alessandra Rizzardi, Luigi Alfredo Grieco, Alberto CoenPorisini. "Security privacy & trust in internet of things: the road ahead computer networks". In: *Comput. Networks*, vol. 76, (2015): 146–164. https://doi.org/10.1016/j.comnet.2014.11.008
14. M. Turkanovic, B. Brumen, M. Holbl. "A novel user authentication and key agreement scheme for heterogeneous ad hoc wireless sensor networks based on the internet of things notion". In: *Ad Hoc Networks*, vol. 20, (2014): 96–112. doi: https://doi.org/10.1016/j.adhoc.2014.03.009
15. N. Ye, Y. Zhu, R.-C. b. Wang, R. Malekian, Q.-M. Lin. "An efficient authentication and access control scheme for perception layer of internet of things". In: *Appl. Math. Inf. Sci.*, vol. 8, no. 4, (2014): 1617–1624. doi: 10.12785/amis/08041
16. A. Alcaide, E. Palomar, J. Montero-Castillo, A. Ribagorda. "Anonymous authentication for privacy-preserving IoT target driven applications". In: *Comput. Secur.*, vol. 37, (2013): 111–123. doi: https://doi.org/10.1016/j.cose.2013.05.007

17. M. Ali, M. ElTabakh, C. NitaRotaru. "FT-RC4: A robust security mechanism for data stream systems". In: *Tech. Rep.* TR-05-024, (2005): 1–10.
18. M. A. Hammad, M. J. Franklin, W. Aref, A. K. Elmagarmid, 2003. Scheduling for shared window joins over data streams. In: *Proceedings of the 29th International Conference on Very Large Data Bases (VLDB).* pp. 297–308. doi: 10.5555/1315451.131547
19. K. Jaswal, T. Choudhury, R. L. Chhokar, S. R. Singh, 2017. Securing the Internet of Things: A proposed framework. In: *2017 International Conference on Computing, Communication and Automation (ICCCA).* pp. 1277–1281. doi: 10.1109/CCAA.2017.8230015
20. Karnouskos Spiess, Savio Guinard, Souza Baecker, et al. "SOA-Based Integration of the Internet of Things in Enterprise Services". In: *2009 IEEE International Conference on Web Services,* (2009): 968–975. doi: 10.1109/ICWS.2009.98
21. Jacob Sakhnini, Hadis Karimipour, Ali Dehghantanha, Reza M. Parizi, Gautam Srivastava. "Security aspects of internet of things aided smart grids: A bibliometric survey". In: *Internet of Things,* (2020): 1–45.
22. T. Tran, O. Shin, J. Lee, 2013. Detection of replay attacks in smart grid systems. In: *2013 International Conference on Computing, Management and Telecommunications (ComManTel).* pp. 298–302. doi: 10.1109/ComManTel.2013.6482409
23. P. Eder-Neuhauser, T. Zseby, J. Fabini. "Malware propagation in smart grid networks: metrics, simulation and comparison of three malware". In: *J. Comput. Virol. Hacking Tech.* (2015): 109–125. doi: 10.1007/s11416-018-0325-y
24. P. Eder-Neuhauser, T. Zseby, J. Fabini, "Malware propagation in smart grid mono-cultures". In: *E & I Elektrotechnik und Informationstechnik,* vol. 135, no. 3, (2018): 264–269. doi: 10.1007/s00502-018-0616-5
25. J. Tian, B. Wang, X. Li, "Data-driven and low-sparsity false data injection attacks in smart grid". In: *Secur Commun Netw,* vol. 2018, (2018): 1–11. doi: 10.1155/2018/8045909
26. X. Liu, Z. Li, "False data attack models, impact analyses and defense strategies in the electricity grid". In: *Electric. J.,* vol. 30, no. 4, (2017): 35–42. doi: https://doi.org/10.1016/j.tej.2017.04.001
27. V. A. Patel, V. K. Patel, G. Panchal. "Integration of IOT and Cloud computing and its issues: a survey". In: *Int. J. Eng. Technol. Sci. Res.,* vol 4, no. 10, (2017): 903–907.
28. L. Yang, C. Ding, M. Wu, K. Wang, "Robust detection of false data injection attacks for data aggregation in an Internet of Things-based environmental surveillance". *Comput. Netw.,* 129, (2017): 410–428. https://doi.org/10.1016/j.comnet.2017.05.027
29. L. Shi, Y. Wang, "Secure data delivery with linear network coding for multiple multicasts with multiple streams in internet of things". In: *Secur. Commun. Netw.,* 2018 (2018): 1–13. doi: 10.1155/2018/3729046
30. E. D. Chowdary, D. Yakobu, 2016. Cloud of Things (CoT) integration challenges. *In: 2016 IEEE International Conference on Computational Intelligence and Computing Research (ICCIC).* pp. 1–5. doi: 10.1109/ICCIC.2016.7919553
31. C. Pu, S. Hajjar, 2018. Mitigating Forwarding Misbehaviors in RPL-Based Low Power and Lossy Networks. *In: 2018 15th IEEE Annual Consumer Communications Networking Conference (CCNC).* pp. 1–6. doi: 10.1109/CCNC.2018.8319164
32. Thomas Renner, Andreas Kliem, and Odej Kao, 2014. The Device Cloud - Applying Cloud Computing Concepts to the Internet of Things. *In: 2014 IEEE 11th Intl Conf on Ubiquitous Intelligence & Computing.* pp. 396–401. doi: 10.1109/UIC-ATC-ScalCom.2014.106.
33. Hamed Haddad Pajouh, Ali Dehghantanha, Reza M. Parizi, Mohammed Aledhari, Hadis Karimipour. "A survey on internet of things security: Requirements, challenges, and solutions." In: *Internet of Things,* (2019). pp. 100-129, doi: https://doi.org/10.1016/j.iot.2019.100129

34. Z. Shelby, K. Hartke, and C. Bormann, "The Constrained Application Protocol (CoAP)", RFC 7252, (June 2014). doi: 10.17487/RFC7252. <https://www.rfc-editor.org/info/rfc7252>.

35. D. Wu, G. Hu, 2008. Research and improve on secure routing protocols in wireless sensor networks. *In: 4th IEEE International Conference on Circuits and Systems for Communications 2008. ICCSC 2008.* pp. 853–856. doi: 10.1109/ICCSC.2008.186

36. D. Conzon, T. Bolognesi, P. Brizzi, A. Lotito, R. Tomasi, M. A. Spirito, 2012. The Virtus Middleware: An Xmpp Based Architecture for Secure IoT Communications. *In: 2012 21st International Conference on Computer Communications and Networks (ICCCN).* pp. 1–6. doi: 10.1109/ICCCN.2012.6289309.

37. H. Kim, 2008. Protection against Packet Fragmentation Attacks at 6lowpan Adaptation Layer. *In: 2008 International Conference on Convergence and Hybrid Information Technology,* pp. 796–801. doi: 10.1109/ICHIT.2008.261

38. R. Hummen, J. Hiller, H. Wirtz, M. Henze, H. Shafagh, K. Wehrle, 2013. 6LoWPAN Fragmentation Attacks and Mitigation Mechanisms. *In: Proceedings of the Sixth ACM Conference on Security and Privacy in Wireless and Mobile Networks.* pp. 55–66. https://doi.org/10.1145/2462096.2462107

39. B. Park. "Threats and security analysis for enhanced secure neighbor discovery protocol (SEND) of IPv6 NDP security." In: *Int. J. Control Automat.,* vol. 4, no. 4 (2011): 179–184.

40. A. Dvir, T. Holczer, L. Buttyan, 2011. "VeRA—Version number and rank authentication in RPL". In: *2011 IEEE Eighth International Conference on Mobile Ad-Hoc and Sensor Systems,* pp. 709–714. doi: 10.1109/MASS.2011.76

3 Role of Cryptography and Its Challenges in Integrating Secured IoT Products

D. Citharthan
Christ the King Engineering College
Coimbatore, Tamil Nadu, India

M. Varatharaj
Kathir College of Engineering
Coimbatore, Tamil Nadu, India

D. Palanivel Rajan
CMR Engineering College
Hyderabad, Telangana, India

CONTENTS

3.1 INTRODUCTION

Cryptology is a study of encryption and decryption. Encryption is a technique used to convert plain (readable) text to cipher (unreadable) text and, decryption is a technique used to convert cipher (unreadable) text to plain (readable) text. There are two methods of encryption, namely symmetric encryption and asymmetric encryption. Symmetric encryption uses symmetric keys for encryption as well as decryption. Asymmetric encryption uses two independent (asymmetric) keys, namely public key and private key. In asymmetric encryption, any one of the keys will be used for encryption, and another key will be used for decryption. Before sending any plain text to the IoT device, it will be encrypted using a symmetric or asymmetric key and converted to cipher text in the sender side. On the receiver side, the cipher text is decrypted using the same symmetric key or asymmetric key pair and converted to plain text. Symmetric encryption is faster than asymmetric encryption. But the problem is that symmetric keys have to store securely, and secure channel is required to transfer the symmetric key. So, combining the symmetric and asymmetric encryption methods will be more efficient in IoT product [1]. Asymmetric encryption is used in the key deployment and can be used to encrypt communication. Symmetric encryption is used where performance is required or where the data has to transmit faster.

3.2 ATTACKS IN NETWORK

According to the Attack landscape H1 2019 report, the number of network attack events measured between January and June was twelve times higher when compared to the same period in 2018, an increase mostly driven by IoT-related traffic. Attacks in the network will be performed by network attackers or unauthorized persons, where they will access the data or modifies the data or even destroys the data. In network, the attacks are classified into the passive attacks and active attacks. In a passive attack, the data will not be modified, whereas, in most of the active attacks, the data will be modified during transmission. There are two techniques in passive attack, namely "release of the contents" where the third-party will access the data sent between the sender and the receiver without any modifications and 'traffic analysis' where the third-party observes the traffic flow between the sender and the receiver. Here data will not be modified, and so the above techniques are called eavesdropping techniques. There are four types of active attacks, namely masquerade attack, replay attack, data modification, and denial of service. In the masquerade attack, the third-party/unauthorized person will send the data to the receiver in the name of the sender. In a replay attack, the third-party/unauthorized person will access the data and again sends the same data to the receiver. In data modification, the third-party/unauthorized person will access the data and sends the modified data to the receiver. In denial of service, the third-party/unauthorized person will deny the service to the authorized person. To avoid the attacks in network, it is necessary to provide network security services for any IoT product.

3.3   NETWORK SECURITY SERVICES

To protect the data from active attacks and passive attacks, it is necessary to provide network security services. There are four types of network security services, namely:

  i. **Access Control**—It prevents unauthorized access to resources.
 ii. **Authentication**—It verifies user identity.
iii. **Confidentiality**—It provides security to the data that has to be sent through the network.
 iv. **Integrity**—It prevents data modification during the data transmission.
  v. **Non-Repudiation**—It prevents from denying of authenticity and integrity of the data.

3.4   SECURITY FOR IoT DEVICE USING CRYPTOGRAPHIC KEYS

Cryptography plays a major role in IoT secure world. In the context of IoT world, cryptography is a technique, which allows us to protect information and assign it to a person or identity. Consider an industry named as XYZ wants to design an IoT product and need to ensure the following security things:

- Check the identity of the IoT product before sending any information to ensure authenticity.
- During the transmission, the information is secure and should not be modified to ensure data confidentiality and integrity.

To ensure the above security things, first, the XYZ industry has to accomplish the following process using cryptographic keys to check the identity of IoT product before sending information:

- Key deployment into the IoT product.
- Process of authentication.

Then, the XYZ industry has to accomplish the following process using cryptographic keys to have a secure and unmodified transmission of information:

- Secure transmission of symmetric a secret key.
- Accomplishing the secure transmission of data.
- HASHing for unmodified transmission of data.
- Signature generation.

3.4.1 Key Deployment into the IoT Product

The steps for key deployment into the IoT product are as follows:

Step 1: First, the XYZ industry has to generate asymmetric public (say, key No. 1.1) and private (say, key No 1.2) key pair in a secure environment using random number generator and asymmetric key pair generator.

Step 2: Next, the XYZ industry has to create a certificate which includes the public key (key No 1.1). This is the formalised way of allowing someone to read the public key (key No 1.1).

Step 3: Using secure deploy the certificate (with public key- key No 1.1) is to be provisioned into the internal memory of the IoT product.

Step 4: Finally, using secure deploy the private key (key No 1.2) is to be provisioned into the secret internal memory of the IoT product.

3.4.2 Process of Authentication

The steps for the authentication process are as follows

Step 1: To check the identity of IoT product, the XYZ industry has to create and transmit random generated raw message (without encryption) to the IoT product.

Step 2: IoT product encrypts the message with private key (key No. 1.2) that has already been deployed in it and send the encrypted message back to the XYZ industry.

Step 3: Using public key (key No. 1.1), the XYZ industry can able to decrypt the message, and if that decrypted message was same as the message created then it ensures that particular IoT product is authenticated and it is having the private key (key No. 1.2) in the internal memory of the IoT product.

3.4.3 Secure Transmission of Symmetric Secret Key

The steps for the secure transmission of a symmetric secret key are as follows:

Step 1: Before sending the information, the XYZ industry has to generate another ephemeral symmetric secret key pair or session key pair (say, key No. 2.1 and key No. 2.2).

Step 2: One of the ephemeral symmetric secret keys (say, key No. 2.1) will be encrypted using the public key (key No. 1.1) and transmitted safely over any transport medium.

Step 3: This session key (key No. 2.1) can be decrypted only by the XYZ industry IoT product, which is having the private key (key No. 1.2). Once decrypted, the session key (key No. 2.1) is loaded into the secured memory of the XYZ IoT product.

3.4.4   ACCOMPLISHING THE SECURE TRANSMISSION OF DATA

The data from the XYZ industry can now be encrypted using the ephemeral secret key (key No. 2.2) and transmitted over any transport medium. This encrypted data can be decrypted only by the XYZ industry IoT product which is having the ephemeral secret key (key No. 2.2). Once decrypted, the data can be loaded into the unsecured memory of the XYZ IoT product and thus accomplishing the secure transmission of data.

3.4.5   HASHING FOR UNMODIFIED TRANSMISSION OF DATA

HASHing is to confirm the data from the XYZ industry has not been modified. This ensures integrity. The arbitrary length of data from the XYZ industry is to be pushed into a HASHing algorithm (example: SHA-256), which will create a unique 256 bit code. Then, the data that has to send from the XYZ industry is transmitted with the accompanying 256 bit HASH value. In the receiver side, HASH SHA-256 algorithm is used to generate HASH code from the data. If the HASH code generated on the receiver side is the same as the HASH code from the XYZ industry, then it is proven that the data during the transmission is unmodified. Figure 3.1 illustrates the steps for HASHing.

3.4.6   SIGNATURE GENERATION FOR UNMODIFIED
TRANSMISSION OF DATA AND AUTHENTICATION

Signature generation incorporates both HASHing and encryption to confirm the data of the XYZ industry has not been modified and has been sent from authorized source. This ensures integrity and authenticity. Before sending the information, the XYZ industry has to generate another asymmetric signing key pair namely signing private key (say, key No. 3.1) and signing public key (say, key No. 3.2). The arbitrary length of data from the XYZ industry is to be pushed into a HASHing algorithm (example: SHA-256), which will create a unique 256 bit code. This unique 256 bit code is encrypted using an asymmetric signing private key (key No. 3.1) to create the encrypted HASH code called a signature. This signature will be appended to the transmitted data. In the IoT product, data is again pushed into a HASHing algorithm (example: SHA-256) to create HASH code. Then, signature has to be decrypted using an asymmetric public key (key No. 3.2), which has to be provisioned into the IoT product to create HASH code. If both the hash codes are same, then it is proven that the data during the transmission is unmodified, and the data has been sent from authorized the XYZ industry, which is only having the asymmetric signing private key (key No. 3.1). Figure 3.2 illustrates the steps for signature generation. Due to large real numbers, problems may arise due to computational overhead which may be reduced by lightweight Shortened Complex Digital Signature Algorithm (SDSA) [2].

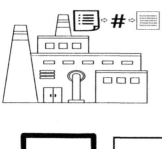

Step 1:

The data that has to send from XYZ industry to the IoT product is pushed into a HASHing algorithm and generate 256 bit code

Step 2:

The data that has to send to the IoT product is then transmitted with the accompanying 256 bit HASH code

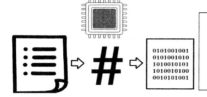

Step 3:

The data in the IoT product is pushed into a HASHing algorithm and generate 256 bit code.

Step 4:

Check whether the HASH code generated in the IoT product is same as the HASH code received from the XYZ industry. If same, then it ensures that data from the XYZ industry was not tampered

FIGURE 3.1   Steps for HASHing.

3.5   CRYPTOGRAPHIC ALGORITHMS

Confidentiality provides network security. Security to the data can be provided by using symmetric keys or asymmetric key pair (public key and private key). If the user uses symmetric keys for providing security, then data encryption standard (DES) algorithm or advanced encryption standard (AES) algorithm can be used. If the user uses an asymmetric key pair for providing security, then RSA (Rivest, Shamir, Adleman) algorithm can be used. Providing authentication and integrity needs fixed-length value, which needs to be appended to the plain text. For calculating the fixed-length value, SHA 256 algorithm can be used. Integrity and authenticity can be provided by using a signature generation. Digital signature algorithm (DSA) can be used in signature generation. Usually, one or more above algorithms are combined to form the desired security property. Depending on the protocols, a grouping of these algorithms is done, and they are called cipher suites.

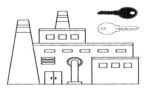

Step 1:

XYZ industry generate asymmetric public key
(key No 3.1) and private key (key No 3.2) in a secure
facility

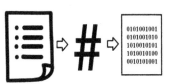

Step 2:

The data that has to send to the IoT product is pushed into
a HASHing algorithm and generate 256 bit code

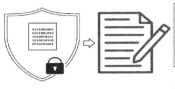

Step 3:

Encrypt the 256 bit code using the private key
(key No – 3.2) and creates signature

Step 4:

The data that has to send to the IoT product is then
transmitted with the accompanying signature

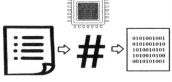

Step 5:

The data received in the IoT product is passed through a
HASH algorithm to generate 256 bit HASH code (1)

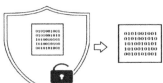

Step 6:

Decrypt the signature using the public key (key No 3.1)
which is to be deployed in the IoT product and, generate
256 bit HASH code (2).

Step 7:

Check whether the HASH code (1) & HASH code (2) are
same. If same, then it ensures that data is originated only
from XYZ industry and has not been modified

FIGURE 3.2 Steps for signature generation.

3.6   RANDOM NUMBER GENERATOR

Random number generators are used to create random numbers that can be used as a symmetric key or asymmetric key pair. Types of random number generators (RNG) are deterministic RNG and non-deterministic RNG. Deterministic RNG will always produce the same output for a single set of inputs. Non-deterministic RNG will generate random data from very random physical elements, such as circuit noise and other low bias sources, and it should be insensitive to deterministic influences, such as power supply noise and cross talk [3]. Both RNGs require inputs called seeds. Seeds also should be highly random, which has to come from high entropy sources. The entropy sources are to be protected from tampering or any other type of malfunction. Direct amplification, oscillator sampling, and discrete-time chaos are some of the IC compatible methods by which it is possible to produce random sequences. Usually, RNG requires noise sources such as thermal noise or shot noise or phase noise (which is actually a byproduct MOSFET thermal noise)

3.7   PUBLIC KEY INFRASTRUCTURE (PKI)

A PKI is a set of hardware, software, people, policies, and procedures needed to create, manage, store, and revoke digital certificates [4]. Actually, the public key in asymmetric key pairs is made public in the form of certificates. Using the private key, the certificate is signed and issued by the certificate authorities. A certificate has a complex chain of trust mechanism that validates it via certificate authorities. Authorities involve in certificate creation, and validations are registration authority and certificate authority.

3.7.1   REGISTRATION AND CERTIFICATE AUTHORITY

Registration authority checks on users to verify their identity using a specific process called certificate practise statement (CPS). Then, it can generate public and private key pair. The generation of keys is optional because registration authority wants to generate keys only if it is not done in a secure personal environment of an industry. Finally, it can instruct certificate authority to issue certificate. A certificate authority is an entity that issues security certificates and public keys, once permission has been received from a registration authority.

3.7.2   REGISTRATION PROCESS

Consider the XYZ industry wants to get a certificate for their IoT product. Then, the steps which have to be followed by the XYZ industry are given below:

Step 1: The XYZ industry requests a certificate from the registration authority.
Step 2: The public and private key pair should be either securely generated by the XYZ industry or by the registration authority.
Step 3: The registration submits a certificate request to the certificate authority.

Step 4: The certificate authority verifies the request and generates the certificate and publishes it.

Step 5: The generated certificate is delivered via the registration authority to the XYZ industry.

3.7.3   REVOCATION PROCESS

Consider the same the XYZ industry wants to revoke the certificate because of lost/stolen private key or termination of service for their IoT product. Then, the steps which have to be followed by the XYZ industry are given below:

Step 1: The XYZ industry requests a revocation to the registration authority.

Step 2: The authorization of the revocation request is checked at the registration authority.

Step 3: The registration authority informs the certificate authority.

Step 4: The certificate authority invalidates the certificate by signing a new certificate revocation list.

3.7.4   CERTIFICATE FORMAT

Certificates structure is formatted as defined by the X.509 standard. Certificates are encoded in order to ease transmission through systems. Encoding of certificates can be done by distinguished encoding rules (DER), and privacy enhanced mail (PEM). In DER, certificates are encoded in binary form, and it will be challenging to transmit through some systems. In PEM, certificates are encoded into text, and it will be easier to send through systems. The encoding can be cumbersome for small embedded devices. So, a simplified version is stored in the embedded device, and in some cases, this is reduced to the raw MCUs public key. A certificate that has to be used by the industry needs to be decoded in order to access the HASH algorithm type, public key, and signature HASH code. The certificate has a complex chain of trust mechanism that validates it via certificate authorities. A certificate contains information about the microcontroller and importantly, a signature. The signature has been encrypted by the certificate authorities' private key. The signature can, therefore, be decrypted by the certificate authorities' public key.

3.7.5   SIGNATURE CONSTRUCTION IN CERTIFICATE AND CERTIFICATE VERIFICATION

Signature construction in the certificate will be done by the following two steps, and they are shown in Figure 3.3:

- Certificate information is taken to convert it into a unique fixed data value by HASH function.
- Then the unique fixed data value is encrypted using certificate authorities' private key and creates signature.

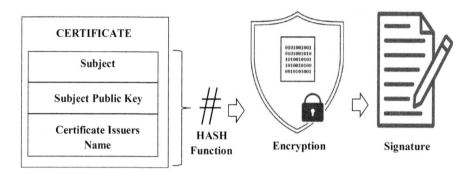

FIGURE 3.3   Process of signature construction in the certificate.

Verification of the certificate by the user can be done by the following steps:

- Take the certificate information and push it into the HASH function to convert it into a unique fixed data value.
- Decrypt the signature using the public key of the certificate authority and get the fixed data value.
- If both the above data values are the same, then it ensures that the certificate has been validated by the certificate authority.

3.7.6   CHAIN OF TRUST

Certificate has been issued by a certificate authority, which forms part of the chain of trust. The security of IoT product can be improved by the chain of trust, which includes product certificate, intermediate certificate, and root certificate. Root certificate private key has to be protected in hardware security module (HSM). The root certificate private key will be used to create and revoke intermediate certificate and not the device certificate [5]. So, even if the unauthorized person accesses the product private key, he/she can only attack the devices associated with the intermediate certificate. IoT product private key (say, X) will provisioned into the IoT product using secure deploy. IoT product will has certificate by which it is possible to get access of public key (say, Y). This IoT product certificate can be verified by verifying the signature using the intermediate certificate public key (say, B). Since the signature in the IoT product certificate will be encrypted by the intermediate certificate private key (say, A). This intermediate certificate can be verified by verifying the signature using the root certificate public key (say, D). Since, the signature in the intermediate certificate will be encrypted by the root certificate private key (say, C), thus the security of IoT product can be greatly improved by the chain of trust.

3.8   CHALLENGES OF CRYPTOGRAPHY IN
INTEGRATING SECURED IoT PRODUCT

In 1994, Peter Shor proved that quantum computers were able to solve discrete logarithm problems in an efficient manner. This means a quantum computer can break nearly all the current classical asymmetric cryptography. So, the important challenge in cryptography is to design new asymmetric cryptography and to prove that quantum computers cannot break it. Elliptic curve cryptography (ECC) can be an efficient alternative for asymmetric cryptography since ECC with 256 bit key is stronger than asymmetric cryptography with 4096 bit key. So, when compared to asymmetric cryptography, ECC uses smaller keys and provide high security and high speed. But the drawbacks of ECC are their complexity and large attack surface [6]. Next, if the user needs encrypted data to process, which is there in cloud, then it is necessary to download all the encrypted data, and it needs to be decrypted every time. This will be time-consuming for the users to search a single piece of encrypted data in the cloud. This can be overcome by homomorphic encryption technique. Homomorphic encryption technique is used to convert encrypted cipher text to another cipher text [7]. It means it allows computation on cipher text and generates an encrypted results. For example, consider an encrypted cipher text $C = E (K,P)$ in the cloud, using homomorphic encryption, it will be converted to $C' = E (F(K), P)$ where $F(K)$ can be any modification of data K and even cloud provider doesn't know the modifications done to the data. But this type of operation is very slow, and it would take a very long time for even basic operations. So, it is necessary to have an efficient general-purpose homomorphic encryption.

3.9 FUTURE DIRECTIONS OF CRYPTOGRAPHY

Modern cryptography is a study of techniques that can be used for securing digital information, transactions, and distributed computation [8]. Homomorphic encryption, honey encryption, functional encryption, and quantum encryption are the future encryption techniques that can be used in cryptography. Homomorphic encryption is a type of encryption, where it is possible to process the data without having access of it. Honey encryption is a type of encryption, where wrong guesses of the key give the information such that eavesdropper thinks it's right but actually not. The main advantage of honey encryption, is that it completely avoids the brute force attack. Functional encryption is a type of encryption where keys are hidden in software. Quantum encryption is a type where the quantum nature of an atom protects the data. Actually, it is very difficult to predict future adaptive protocols, but it is easy to say that future adaptive protocols should have distributed trust. One such protocol is the blockchain protocol. A public blockchain and private blockchain are the two types of blockchain [9]. Public blockchain is a permissionless, whereas private blockchain is a permissioned one. Usually, blockchain uses cryptographic HASHes buildup over time. A Merkle tree represents an interesting application of HASH function where they are said to form a peer to peer network. So, it ensures that the data blocks received from other peers are unaltered and undamaged. For example, consider a distributed control system where IoT controller wants to control the actuator. Nodes on the network will work together to command the action and actuator to perform the action. Blockchain will also remove authorizations from the improper sensors in the network. For example, if the controller receives the data from many sensors and if any one of the sensors begins to provide data, which is not within the threshold values, then the controller can update the blockchain to

TABLE 3.1

Steps in Secured Data Transmission and Its Cryptographic Services

Steps in Secured Data Transmission	Cryptographic Services
Key deployment into the IoT product	Encryption and decryption
Authentication process	Encryption and decryption
Secure transmission of a symmetric secret key	Encryption and decryption
Secure transmission of data	Encryption and decryption
Ensuring unmodified transmission of data	HASHing
Ensuring unmodified transmission of data and authentication	Digital signature

remove authorizations from that particular incorrect sensor [10]. Usually, updated blockchain will be HASHed with other updated HASHes through a Merkle tree. Using distributed ledgers as intermediaries, a platform called Hermes can be used for trading sensor data against malicious behaviour [11–14].

3.10 SUMMARY

In this chapter, implementation of data transmission in secured IoT product was discussed in which key deployment into the IoT product, authentication process, secure transmission of a symmetric secret key, secure transmission of data, HASHing for unmodified transmission of data, and signature generation were explained. Table 3.1 shows the various steps for the implementation of data transmission in the secured IoT product and its cryptographic services.

We also explained the Public Key Infrastructure (PKI) and its chain of trust with an example. Apart from that, we discussed the challenges of cryptography and the need for efficient asymmetric cryptography, elliptic curve cryptography, and homomorphic encryption. Finally, we discussed the future directions of cryptography in integrating secured IoT products, where it ensures the importance of distributed trust using a blockchain protocol.

REFERENCES

1. C. H Meyer, "Cryptography—a state of the art review", Proceedings, VLSI & Computer Peripherals, August 2002.
2. Muhammad Arif Mughal, Xiong Luo, Ata ullah, Subhan ullah, Zahid Mahmood, "A Lightweight digital signature based security scheme for human-centered internet of things", *IEEE Access*, VOL 6, June 2018.
3. Craig S. Petrie, J. Alvin Connely, "A noise based IC random number generator for applications in cryptography", *IEEE Transactions on Circuits and Systems: Fundamental Theory and Applications*, Vol. 47, No. 5, May 2000.
4. Brian Russel, Drew Van Duren, *Practical Internet of Things Security, A practical, indispensable security guide that will navigate you through complex realm of security building and deploying systems in our IoT-connected world*, PACKT Publishing, June 2016.

5. Hayden Povey, Founder & CTO-Secure Thingz, *Secure your Application from Design to Deployment*, November 2017.
6. Skilton, Mark, and Felix Hovsepian. *The 4th industrial revolution: Responding to the impact of artificial intelligence on business*. Springer, 2017.
7. Moncef Amara, Amar Siad, "Elliptic curve cryptography and its applications", 7th International workshop on systems, signal processing, and their applications, 2011
8. Jean-Philippe Aumasson, *Serious Cryptography, A practical introduction to modern encryption*, No Strach Press, 2018.
9. Jonathan Katz, Yahuda lindell, "Introduction to modern cryptography", *Chapman & Hall/CRC*, Taylor & Francis Group, 2008
10. Fahad Alkuridi, Ibrahim Elegendi, Kumudu S. Munasinghe, Dharmendra Sharma, Abbas Jamalipour, "Blockchain in IoT Security: asurvey", *28th International Telecommunication Networks and Applications Conference (ITNAC)*, November 2018.
11. Shitang Yu, Kun Lv, Zhou Shau, Yingcheng Guo, Jun Zou, Bo Zhang, "A high performance blockchain platform for intelligent devices", *1st IEEE International conference on hot information-centric networking*, January 2019.
12. Pavlos Tzianos, Georgios Pipelidis, Nikos Tsiamitros, "Hermes: an open and transparent market place for IoT sensor data over distributed ledgers", *IEEE International conference on blockchain and cryptocurrency (ICBC)*, July 2019.
13. P. Karthikeyyan, and S. Velliangiri. "Review of Blockchain based IoT application and its security issues." In *2019 2nd International Conference on Intelligent Computing, Instrumentation and Control Technologies (ICICICT)*, vol. 1, pp. 6–11. IEEE, 2019.
14. S. Velliangiri, J. Premalatha, Intrusion detection of distributed denial of service attack in cloud. *Cluster Computing* 22, 10615–10623, 2019.
15. S. Velliangiri, R. Sekar, and P. Anbhazhagan, "Using MLPA for smart mushroom farm monitoring system based on IoT", *International Journal on Networking and Virtual Organisations*, Vol. 22, No. 4, pp. 334–346, 2020.

4 Blockchain-Based Security for IoT in Cloud—A Review

L. Sherly Puspha Annabel and Kripa Sekaran
St. Joseph's College of Engineering
Chennai, Tamil Nadu, India

CONTENTS

4.1   INTRODUCTION

The Internet of Things (IoT) is an evolving technology that has immensely provided a lot of scope in engineering and science domains, making the work of human-human workforce easier. It emphasizes mostly a new generation workforce, increasing the interaction between humans and machines in a broader way. The IoT has now become an inherent feature of various applications of modern-day living. Therefore, we can strongly infer that IoT is an amalgamation of multiple technologies that support people's needs. Some of the examples include information technology, actuator technology, and advancements in analytics.

The integration of diversified techniques results in complexity when applied to a larger application platform. Integration of devices, interconnection of networks, and distributed environment in IoT emphasize the need for a central server that provides authentication and higher processing. In this current scenario, there is a chance of unreliable interconnections between the devices that results in illegal authentication and spoofing of devices.

This scenario emphasizes the fact that IoT would tend to be more complicated by linking it to a network of multiple things (NMT), for providing sufficient digital access. The NMT devices collect a voluminous amount of information from the basic environment in which they survive. These NMT devices have an interaction with the network through the internet, which highlights the role of central server storage. These rich interactions pave the way for an enormous amount of data generation and supporting dependable services through specialized data management servers (SDM). In this case, providing steadfastness and dependability in various services is not sufficiently secure, resulting in a breach of privacy and security issues across the network. Moreover, many provisions are specified for revealing the subtle aspects of the data to the external world in the form of spoofing and unproved authentications leading to the various concerns in IoT privacy and that are difficult to encounter.

To overcome the issues of NMT, centralized maintenance of data produced by NMT can be eliminated and blockchain (BC) technology can be introduced. This chapter highlights the application of BC technology in IoT and security in the cloud by analyzing the various security concerns addressed during component interaction.

4.2 TERMINOLOGIES

4.2.1 INTERNET OF THINGS

The ever-increasing outgrowth of the internet has diversified the exchange of information, making IoT to present new opportunities that provide users a definite advantage over others. The reliability of IoT is because of convenience for the people to use due to the strong presence of a device network. Inefficiencies are caused due to improper communication, but when a network of devices is used, proper communication is ensured by the smooth transfer of data packets saving appropriate time and money.

IoT saves time and money along with relevant technological advancements resulting in increased quality of services of the assigned tasks [1]. A variant of IoT called Platform IoT uses sensors to collect relevant information and involves in the transmission and reception of the data to the server through appropriate connectivity. IoT's software part focuses on making decisions with the data and builds a user interface to enable system-user interaction.

4.2.2 BLOCKCHAIN

BC encompasses two segments—Blocks, which has a collection of transactions in line with hash values and the other segment chain, which is basically a collection of blocks having hash values of previous sectored blocks. This technology eliminated the need for a central entity to authenticate for transactions between them. Thus auditing is done independently across all the network nodes resulting in robustness to the Byzantine Fault Tolerance problem. A shared digital ledger that is incorruptible is developed, which has the advancements to record all details. Bitcoin is a cryptocurrency introduced by BC technology. The Merkle tree was specifically used for creating a single block from many documents for improving efficiency. Over the years, the trust-worthiness of BC has increased, and its scope is much beyond just crypto-currency [2, 3].

4.2.3 CLOUD COMPUTING

Cloud computing is a novel model that provides an easy-to-use business model for organizations to adopt various information technology practices without a wholesome investment. Cloud computing also creates a new era of internet-based, robust performance oriented collective computing systems in which services are provided by a group of computational resources. Demand network access caters to a set of configurable computing resources that are provisioned with very low management interference. Multi-tenancy and elasticity are the two major features of the cloud. Multi-tenancy enables the distribution of the same service copies with other users. Elasticity focuses on scaling the resources allotted to a service in accordance with the present demands of service. Nevertheless, the onus is upon service and cost availability and effective utilization of resources.

4.3 INTERNET OF THINGS

The IoT is a special type of a collective environment in which all non-living things, people, and other living things have a unique identity and transfer data between them without any visible interaction. It is the amalgamation of multiple technologies, including the internet and various wireless communication technologies [4]. A specific thing which is represented in the IoT scenario can be a human-designed object, a human being with an implant to monitor heart, any other living thing say cheetah with a transponder made of biochip or any sensor-enabled car. These unique things are given a single and a separate IP address and have the capability to transfer data to different domains. IoT is also strongly related to communication between different machines in power and major manufacturing industries. The IPv6 has huge address storage, and it can be used to specify a separate address to each object on earth, which is one of the salient features for the development of IoT.

IoT assumption is that interactions occur between humans and objects, resulting in an exchange of a large amount of information. Nowadays, IoT devices are used by multiple organizations to make good business decisions based on real and continuous data, which helps in achieving the satisfaction of customers. A distinct enterprise stores data collected from the IoT, which grows exponentially that paves way for the creation of cloud storage [5]. The cloud environment appears to be a good choice for the storage and retrieval of IoT data. Different companies store the details on site since it is not secure to back them up on the cloud. The cloud environment has the following advantages to ensure the storage of IoT data. Firstly, a straight and distinct connection is provided between the cloud provider and the devices. This straight link helps to store data speedily, and so it requires low cost and storage. Secondly, the management of data and storage capacity is the sole problem of the service provider, and the company uses the required service only.

4.3.1 ARCHITECTURAL ELEMENTS OF IoT

Embedded systems play an important role in developing or design an IoT environment. The IoT system has four key constituents [6]. The first and foremost is the internet. Secondly, a device that can transfer real-time and voluminous data over the network. Thirdly, a gateway enabled network which creates internet protocol from communication protocols. Fourthly, the collected data are stored by the use of back-end services, which contains cloud or enterprise database systems. Building and proper designing of the IoT systems can be a tough task. New hardware-enabled and software systems are being designed as well as developed for IoT systems. Moreover, many user-friendly tools are offered to bring reality in the IoT environment. With the increased pace of development of sensors, real-time data capture has also become an easy one.

4.3.2 EXAMPLE

Most of IoT devices have specific sensors that sense the variations in pressure, different sounds, light, and unique motions. These specific sensors are factory-made through the process of lithography and are a type of circuit planned to complete a distinct task and also combined with a microprocessor and connected to a wireless radio for communication.

4.4 BLOCKCHAIN

BC is a collection of blocks that contains all transaction record that happens in a BC network. Each block has a block-header and block body or otherwise known as a transaction counter. Transaction counter is used to cover all the transactions. The block size plays a major role in deciding the maximum number of transactions. As illustrated in Figure 4.1, a unique block header encompasses the following [7].

- Nonce means any 4-byte number whose value has 0 in the beginning and gradually increases for every transaction hash.
- Block version specifies the version of software and rules of validation.
- A timestamp that represents the current world time.
- Merkle Tree root hash which specifies the transaction hash value and collection of all transactions.
- Parent block hash has the previous block hash value.
- N-Bits that represent the transaction verification bits.

Since all completed transactions are stored in a documented blocks; this technology can be named as a public ledger. This current chain encompassing blocks develops continuously as new and distinct blocks are added. The BC technology has important characteristics of audibility and persistency, which ensures cost-effectiveness and improved efficiency.

4.4.1 BLOCKCHAIN PROPERTIES

4.4.1.1 Working Principle of Blockchain

The following is the working principle of BC:

Step 1: A private key that is used to sign the transactions digitally is applied along with a public key to enable node communication. This signed transaction is communicated and sent by a node that has done the transaction.

Step 2: This transaction is verified and confirmed by all the remaining nodes in the network and discards any invalid transaction. This step is called verification [8].

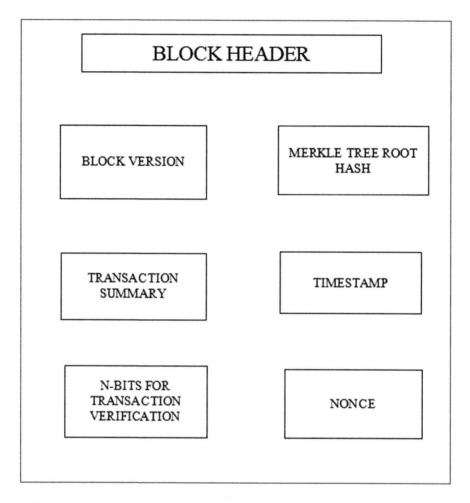

FIGURE 4.1   Components of a block header.

Step 3: The legitimate transaction is converted to a block and mined to get a nonce. This nonce is broadcasted to all the other nodes.

Step 4: A block, which is newly generated is checked for its containment of legal transactions and accuracy by using the relevant hash value. After confirmation, BC adds this newly generated block and updates itself. If it is an unconfirmed block, the block is withdrawn and rejected.

4.4.1.2   Verification

A private key and a public key remove the issues of duplication in BC technology. The public key is an open key which is visible and known to all the other nodes, whereas a private key is of a secretive nature and unknown to other nodes. The encrypted transaction is sent across all the nodes and these nodes decrypt the transaction using their respective keys with a public key. Verification of the signature ensures that the initial node is not modified.

4.4.1.3 Characteristics of Blockchain

The following Table 4.1 lists out the various the characteristics of BC.

4.4.2 APPLICATIONS OF BC

The first and foremost application of BC is in the field of agriculture that includes soil data, crop data, manure data processing, and marketing the data related to agro-products, seeds, etc. It is used to maintain, transport, store the records, and process the transaction of digital files of industries in the business sector. Distribution of digital currencies, sales of used goods is also achieved. It is used for the maintenance of energy generation data, energy suppliers, tariff data, resource tracking. It is also used to deal with data related to food delivery, online order, food quality assurance, food packing, etc. It is applied in the health care industry where the patient's health data, prescription data, symptoms, case history are maintained [9]. It also plays an

TABLE 4.1
Characteristics of Blockchain

S. No	Characteristics	Description
1	Persistence	If a node validates a transaction and approves the transaction record, its copy is broadcasted to all the nodes and that record is not roll backed or permanently deleted from the BC.
2	Decentralization	Centralization results in high cost and low performance. BC maintain data integrity and consistency.
3	Security	Two key cryptographic techniques are used for security in BC. While the private key is used to sign the initiated transaction, public key is used to specify the BC network.
4	Anonymity	In BC, public keys are used to interact among nodes across the entire BC network to keep the user identities safe.
5	Resilient back-end	Every distributed node within the BC, maintains a copy of the preserved public ledger. This prevents the network from any possible failures.
6	Scalability	Many number of hash values can be stored due to the presence of a 160 bit address space.
7	Transparent	BC network can accommodate changes and these changes can be viewed publicly by all nodes in a BC.
8	Good Efficiency	Since, third party cannot get involved in the transaction, the time limit required to confirm dealing is optimum and improves efficiency.
9	Smart Contracts	Definition of different policies and access privileges are written as programs using smart contracts.

essential role in the transport and logistics segment where transport records, logistics service, vehicle tracking, toll data, etc. could be maintained. This is also used in the maintenance of cryptocurrency, currency exchange, money deposit, and transfer, social banking, etc.

4.5 CLOUD COMPUTING

It is defined as the service provided in accordance with the demand made by the users. Information provided by the users is stored in the network of data centers and not in a local server [10, 11]. It is a pay and uses service where the user pays for the type of services they require. Users can access their files from different locations and improve the need for having cost-effective hardware.

Some of its salient features include broader network access (BNA), elasticity, and allocation of resources. BNA helps to access data from various locations for the user by using different computing resources. Moreover, the resources are allocated in accordance with the client's demands. Elasticity is one other feature by which the resources can be scaled based on the need. Inspite of being an ocean of standard features, the cloud security remains the one for concern. The architecture of cloud plays an important role in its security and privacy concerns. Vulnerabilities and attacks have to be closely monitored for cloud security.

4.5.1 ATTACKS ON CLOUD COMPUTING

Some of the attacks that the cloud computing environment [12] is susceptible to:

- **Denial of service (DoS) attack**: Genuine users are stopped from accessing the cloud services by illicit users. In this attack, these users send irrelevant messages to the network for requesting authentication and stop genuine users from accessing their requested service.
- **Cloud malware injection**: Any malicious virtual machine or any service is injected into the environment, and the malware tricks the new service as a system instance itself.
- **Interception of communication**: Two user communications can be intercepted and exploited.

4.5.2 CLOUD SERVICE ARCHITECTURE

Cloud service architecture has three service architectures. Figure 4.2 illustrates the various components of cloud service architecture.

4.5.2.1 Infrastructure as a Service (IaaS)

There is a cloud service provider (CSP) that facilitates users with an interface, which is virtual in the form of a hardware infrastructure to host the relevant data. Own operating systems (OS) and other applications can be used by the various users to

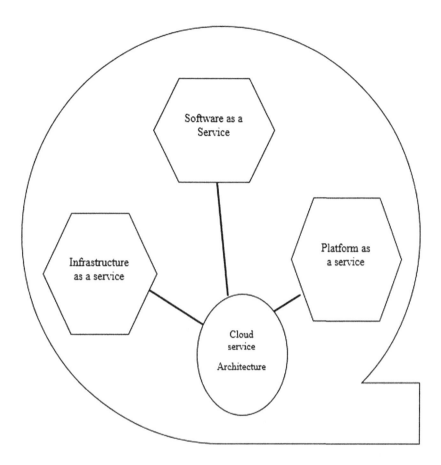

FIGURE 4.2 Components of cloud service architecture.

perform processing in their application. There are various issues of security in this architecture. These issues are listed below:

- Confidential data accessed by unauthenticated users.
- Illegal user causing theft of data [13].
- Cloud applications and activities are not properly monitored.
- Host machines are likely to monitor the main virtual machine.
- Virtual machine's capability to control others machines of same kind.

Security issues solution in this architecture is:

- Network monitoring
- Firewall implementation
- Network segmentation

4.5.2.2 Platform as a Service (PaaS)

This system enables different users to create their applications along with managing them. A base operating system is given to the users, along with the appropriate infrastructure to create applications.

Security issues in PaaS are:

- Secured software development process is not present.
- System failure recovery.
- Service level agreement does not have all provisions.
- Vendors are providing legacy applications.

Security issue solutions in PaaS are:

- Access control policies are sufficiently encapsulated.
- Admission requests enforce authorization.

4.5.2.3 Software as a Service (SaaS)

Cloud vendors provide the necessary infrastructure, applications, and operating systems. It is also referred to as 'software on demand'. Users pay an adequate subscription to use the software and access their services over the internet.

Security issues in this SaaS are:

- Compliance standards are not maintained regularly.
- CSP's operations cannot be accessed.
- Improper authentication and authorization.
- Breach of data.

4.6 SECURITY THREATS AND CHALLENGES IN IoT

The internet engineering task force (IETF) has figured out many IoT security threats [14, 15]. These threats are listed below:

- Cloning by an untrusted source.
- Substitution with lower quality.
- The man in the middle attack.
- Replacement of firmware by an attacker.
- Data privacy threat.
- Attack on routing.
- Eavesdropping on the network.
- Security parameter extraction.

Along with these attacks, the following are some of the security challenges [16].

4.6.1 UNIQUE DEVICE IDENTITY

Each IoT device is given a unique ID. Domain name servers (DNS) give names to the IoT devices that are connected to the network. But the DNS is prone to many kinds of attacks like man-in-the-middle attack.

4.6.2 FIRMWARE

Updation of firmware is a very challenging issue since new security breaches are getting introduced in the webspace. IoT device users have to keep a track record of updates that are installed on these devices. Sometimes, users may have to unmount and install updates and firmware if the device does not support live updates. The introduction of a new device management system may reduce firmware update issues [17].

4.6.3 AUTHORIZATION AND AUTHENTICATION

Sensitive data can be produced by IoT devices, and these are vulnerable to various security attacks. Authentication must be provided to data to transmit them across the gateway. Authorization is as important as authentication since attackers may get read or write access to these sensitive data.

4.6.4 ENORMOUS IOT DEVICE MANAGEMENT

An increase in the quantity of IoT devices in the network results in complicated management of them. New security vulnerabilities are introduced, and no generic management has been updated to overcome them.

4.6.5 SECURITY ALGORITHM IMPLEMENTATION

IoT devices are generally small, along with limited processing and memory capabilities. So the implementation of cryptographic algorithms is difficult, leading to encryption, and decryption difficulties. However, the implementation of certain lightweight algorithms may reduce the possibility of eavesdropping to some extent.

4.6.6 INCIDENT MANAGEMENT

Since IoT can be inserted or kept in any possible thing, failure introduces a considerable problem. A channelized digester recovery plan and a suitable incident management plan have to be designed for real-time IoT devices since the sensors handle very sensitive information.

4.6.7 MANAGEMENT AND DETECTION OF VULNERABILITIES

A challenging issue is a present forefront in the form of detection and management of security vulnerabilities of nodes in IoT. Detection of an affected IoT node is very difficult because of the dense network of IoT devices.

4.6.8 DISRUPTION OF SERVICES

Disruption of services in IoT devices may include compromise of IoT devices, stolen or physically damaged devices. Therefore, the high availability of IoT devices is inevitable for real-time systems monitoring.

4.6.9 DATA INTEGRITY AND PRIVACY

Protection of privacy and integrity is a very challenging issue. Only an authenticated user should have access to sensitive data. Proper authorization from the user is required before access to the data. Secure disposal of data must be done when it is not required.

4.6.10 OTHER HUMAN FACTORS

Lazy user handling of IoT devices is quite a challenging issue. For example, if a damaged machine is not changed by the user, it could result in a life threat for anybody.

4.7 INTEGRATION OF BC WITH IoT TECHNOLOGIES

BC technology would contribute an improved solution to most of the problems met by IoT systems [18–20]. BC is a ledger-based tamper-resistant technology that characterizes an unceasingly preserved and organized database considering rising aspects and composed data sample sets. Here communication occurs through neighbors or peers instead of a central node. All the transactions happen through the BC address, and once when the transactions are entered in the database, it will not be altered, and it permanent]. BC, when integrating with IoT, provides robustness, high reliability, single control authority elimination, economical in terms of infrastructure development, tamper-proof, security, privacy, distributed file sharing facility, and peer to peer message passing facility. Though BC, when it is integrated with IoT, overcomes the various security and privacy issues, it has reasonable limitations that implement this integration of IoT and BC challenging [21, 22].

4.7.1 CHALLENGES INVOLVED IN THE INTEGRATION OF IoT AND BC

The challenges comprise of various limitations that include ledger with limited storage facility, developments in recent technology not as a whole, absence of a skilled workforce, improper legal codes and standards, processing speed and time variations, capabilities of various computing resources, and issues in scalability [23, 24].

4.7.1.1 Storage and Scalability Issues

Some IoT devices may produce an enormous quantity of data, which would make it tough for the integrations of BC with IoT as the existing BC cannot handle such large transactions. Hence, research should be done to resolve these issues before integrating these two technologies. By making use of various devices like embedded and communication devices, IoT could save some reasonable amount of data that is forwarded by IoT to BC. Making careful usage of data, data compression, high bandwidth allocation, and latency reduction could reduce the data transmitted and stored by IoT, thereby improving the integration between IoT and BC.

4.7.1.2 Security

IoT devices have to face lots of security issues as there is high device heterogeneity. Also, properties like mobility, scalability, and communication through wireless mode, attacks on IoT networks affect the security. Another major challenge is data reliability during data communication. IoT devices have to be thoroughly monitored and tested before integration with BC, as the failure of these devices may lead to threats like denial of service and eavesdropping etc. These devices should be placed in an appropriate place to avoid any physical damage. If at all any device failure happens, techniques should be incorporated to report as soon as the fault occurs immediately. Run-time up-gradation and mechanism to reconfigure must be positioned in the IoT to make it run in time as these devices are easily be hacked, and it is difficult for the tools to update one by one. The availability of a Unique Global Identifier and the key pair that is asymmetric in each IoT device of the BC network will provide security for the data [25].

4.7.1.3 Anonymity Along with Data Privacy

As many IoT devices deal with sensitive data, it is necessary to maintain the privacy and anonymity of those data. Secured cryptographic software has to be installed in the methods as these data are not allowed to be used by the public without prior permission. Hence, it is necessary to use devices like gateways and cryptographic hardware to provide cryptographic operations and security. Trust and data integrity are the other two important features that have to be taken into consideration when IoT is integrated with BC. To avoid the overloading of BC with massive data generated by the IoT, the privacy-preserving public auditing mechanism for restricted access control could be done.

4.7.1.4 Smart Contracts

A smart contract is nothing but a pool of functions and also states which exist in a BC address. The IoT devices call functions in a contract. Events will be fired if there is anything that needs to be addressed. If not, the BC would guarantee the transaction and hence, provide reliable and secure data processing. However, smart contracts validation, smart contract overload, decentralized intelligent contracts, becomes a challenge during integration. Therefore, smart contract should be designed in such a way that various constraints of the IoT are taken into consideration. Smart contracts should also relax their distributive nature to resolve the big data present in the cloud environment that is needed by the IoT [26]. An innovative discovery mechanism and actuation mechanism from the smart contracts directly would also make the IoT applications more powerful and faster [27].

4.7.1.5 Legal Issues

Our country's law about data privacy also dramatically affects the IoT sector. The existing laws are not updated and have to be carefully changed, which can simplify the IoT devices security feature certification to build a more secure IoT network.

4.7.1.6 Consensus

Though there are efforts taken to integrate BC in various IoT devices, the process of mining is still a significant challenge as IoT is mostly composed of resource-constrained devices with tremendous processing power. Research work should concentrate much on this area and influence the distributed environment and worldwide potential of the IoT for adapting the consensus in the IoT [28].

4.7.2 Challenges in Implementing Cloud-based IoT and BC Technologies

Although cloud-based IoT has numerous benefits and capable of resolving a wide variety of problems, the challenges still prevail. This section briefly elucidates the several possible concerns that prevail in implementing all these technologies.

4.7.2.1 Security and Privacy Concerns

The first and the foremost concern is about the massive shortage of resources to train upcoming generations about the design of IoT. The absence of well-intimated decisions in IoT cost-benefit analysis and lack of metrics for the identification of security in IoT leads to security issues. Besides, there is no enough evidence on maintainability and upgradeability. As far as privacy is concerned, there are no strict rules for data collection and data usage. There are no multi-party models for enforcing transparency and expression. There is a massive shortage of resources for developing integrated IoT devices with privacy protection. Also, there is no enough protection rendered for the IoT data. The growth of IoT devices linked to the cloud infrastructure may lead to latency due to the farther distance of IoT devices enormous processing of data established from them. An attacker may post a distributed denial of service (DDoS) attack and consume the cloud servers, triggering performance and security issues to the cloud environment [29].

4.7.2.2 Solution to Security Issues

The vital essentials of BC are participant formulated transactions and the blocks that are the recorder versions of those transactions. The correctness of the sequence of transaction details are monitored by the recorder block and do not allow any alteration of the available data. If the recorded data are in sequential order, the necessity for a chain approach begins. These transaction details are shared with the nodes that are participating in the chain, which nullifies the central server concept. Also, BC permits the data to go into the distributed ledger for every transaction that ensures and provides secure authentication. The distributed ledger does not allow any misinterpretation, several wrong authentications of data. This provides the data to flow sensors to the user via a network, router, internet, distributed BC analytics more consistent and secure. Also, BC removes existing IoT single thread communication and improves the privacy and consistency of IoT systems by creating it to be robust. Additionally, the scalable computing power of cloud servers delivers an enhanced data analytics feature for the data, that is, sent from the various IoT devices [30].

4.7.2.3 Interoperability Issues and IoT Standard and Legal Issues

Lack of awareness about the design risk protocols and deficiency of budget and resources lead to various interoperability issues. Fewer efforts are put in for the development of many standards and protocols. There is no documentation for good design practices, and no standard legal policies are followed for the maintenance of IoT devices compatibility. There is not much development about IoT data usage, data sharing. There is no clear cut law or policy or regulation that exists to deal with the liability issues and legal issues. Limited investment is rendered for the research and development of IoT related activities. There are not enough infrastructure resources for strengthening the IoT associated happenings. No industrial and technical progress is made to study and evaluate the outcome of the IoT in the developing countries. There is an urgent necessity to have regulatory coordination about the policy plan to monitor the continuous growth of the IoT in various sectors. Apart from the issues mentioned above, the right issues, economic issues, and developmental issues prevail.

4.7.2.4 Solution to Infrastructure Issues

To handle the enormous quantity of data in big-scale IoT systems, there is a demand for better internet infrastructure, which is taken care of by BC. BC functions permit the IoT systems to keep track of the enormous amount of connected and networked devices. The best way, that is, applied by BC to solve this issue is by providing distributed networks where autonomous device coordination, peer-to-peer networking, and distributed files sharing tasks could be accomplished. The BC well co-ordinates the transactions between various devices.

4.8 FUTURE RESEARCH FOCUS OF IoT AND BC TECHNOLOGIES

As security is alarming issues while the integration of IoT and BC, it is essential that extensive research work has to be carried out to predict the possible threats.

1. Formulation and automation of proper mitigation policy with minimal human intervention should be done.
2. Absence of openly accessible and representative IoT datasets plays an essential role in IoT security research. There should be a standard for facilitating and sharing the datasets for research purposes [31].
3. The enormous power required for processing the public BC networks needs to be addressed.
4. Also, we need to focus on monitoring the threat landscape issues. There are various issues like attackers misuse the security and privacy features,

anti-forensic techniques, thereby avoiding the examination and forensic inquiry attempts and gain access to protected communications between IoT devices [32].

5. Many of the IoT devices would be located and placed in an openly accessible location, which is obviously under the control of adversary, and research work should be undertaken to ensure the security of the data that are saved in those IoT devices.

6. Future research should focus on introducing and developing open-source frameworks to cope up with excessive energy consumption while contributing more operative and capable services.

7. Future research should focus on building a cost-effective method of implementing the BC that provides security and privacy for IoT devices.

8. A unique security solution that could be strong against concerted attacks along with the guarantee of implementation feasibility of results, in case if there is a low resource-constrained IoT devices must be designed.

9. Due to the various volume of resources that are provided, individual security cannot be installed for all the BC IoT architectures, hence, developing a powerful and flexible security framework for BC IoT architectures becomes essential.

10. The design of energy effective consensus protocols becomes necessary as storage limited and resource-constrained IoT devices are not significant enough of meeting the extensive power depletion in BC processing.

4.9   CONCLUSION

For the past two to three decades, the integration of more than two technologies like IoT, cloud computing, BC, BigData, and data mining has recovered people's lives. Out of these technologies, BC, IoT, and cloud computing is the great technological disruptions, and their integration would produce a healthier outcome in every potential area. This chapter presented a survey about the basics of various technologies like IoT, BC, and cloud computing and the integration and scope of these technologies in various applications. Besides, a detailed discussion about IoT related security issues and their remedies were provided. With the development of BC, its integration into both IoT and cloud has evidenced to be advantageous over the centralized systems. Additionally, this chapter delivered a study of the main challenges that BC and IoT must report for them to work together effectively. Planning the solutions into the challenges exposes that many additional mechanisms are required for the implementation of BC in cloud-based IoT systems. This chapter also discussed the future research efforts that have to be taken for the benefits of the BC within diverse, challenging parts of the IoT and cloud. To conclude, we did an in-depth survey of BC, IoT, cloud computing along with their key characteristics, their integration, challenges, solution, future research directions that have the potential to produce major outcomes of various application domains shortly.

REFERENCES

1. Barreto, Luciano, Antonio Celesti, Massimo Villari, Maria Fazio, and Antonio Puliafito. "An authentication model for IoT clouds." In *2015 IEEE/ACM International Conference on Advances in Social Networks Analysis and Mining (ASONAM)*, pp. 1032–1035. IEEE, 2015.
2. Thakore, Riya, Rajkumar Vaghashiya, Chintan Patel, and Nishant Doshi. "Blockchain-based IoT: asurvey." *Procedia Computer Science* 155 (2019): 704–709.
3. Mistry, Ishan, Sudeep Tanwar, Sudhanshu Tyagi, and Neeraj Kumar. "Blockchain for 5G-enabled IoT for industrial automation: asystematic review, solutions, and challenges." *Mechanical Systems and Signal Processing* 135 (2020): 106382.
4. Hwang, Yong Ho. "IoT security & privacy: threats and challenges." In Proceedings of the 1st ACM Workshop on IoT Privacy, Trust, and Security (2015), pp. 1–1.
5. Biswas, Sujit, Kashif Sharif, Fan Li, Boubakr Nour, and Yu Wang. "A scalable blockchain framework for secure transactions in IoT." *IEEE Internet of Things Journal* 6, no. 3 (2018): 4650–4659.
6. Bokefode, Jayant D., Avdhut S. Bhise, Prajakta A. Satarkar, and Dattatray G. Modani. "Developing a secure cloud storage system for storing IoT data by applying role-based encryption." *Procedia Computer Science* 89 (2016): 43–50.
7. Sultan, Abid, Muhammad Azhar Mushtaq, and Muhammad Abubakar. "IoT security issues via blockchain: areview paper." In Proceedings of the 2019 International Conference on Blockchain Technology, pp. 60–65. 2019.
8. Kumar, Nallapaneni Manoj, and Pradeep Kumar Mallick. "Blockchain technology for security issues and challenges in IoT." *Procedia Computer Science* 132 (2018): 1815–1823.
9. Alkhalil, Adel, and Rabie A. Ramadan. "IoT data provenance implementation challenges." *Procedia Computer Science* 109 (2017): 1134–1139.
10. Hussein, Nidal Hassan, and Ahmed Khalid. "A survey of cloud computing security challenges and solutions." *International Journal of Computer Science and Information Security* 14, no. 1 (2016): 52.
11. Sadique, Kazi Masum, Rahim Rahmani, and Paul Johannesson. "Towards security on internet of things: applications and challenges in technology." *Procedia Computer Science* 141 (2018): 199–206.
12. Singh, Ajey, and Maneesh Shrivastava. "Overview of attacks on cloud computing." *International Journal of Engineering and Innovative Technology (IJEIT)* 1, no. 4 (2012).
13. Parikh, Shalin, Dharmin Dave, Reema Patel, and Nishant Doshi. "Security and privacy issues in cloud, fog and edge computing." *Procedia Computer Science* 160 (2019): 734–739.
14. Chakraborty, Rishi Broto, Manjusha Pandey, and Siddharth Swarup Rautaray. "Managing computation load on a blockchain-based multi-layered Internet-of-Things network." *Procedia Computer Science* 132 (2018): 469–476.
15. Ray, Partha Pratim. "A survey of IoT cloud platforms." *Future Computing and Informatics Journal* 1, no. 1-2 (2016): 35–46.
16. Sfar, Arbia Riahi, Enrico Natalizio, Yacine Challal, and Zied Chtourou. "A roadmap for security challenges in the Internet of Things." *Digital Communications and Networks* 4, no. 2 (2018): 118–137.
17. Banerjee, Mandrita, Junghee Lee, and Kim-Kwang Raymond Choo. "A blockchain future for internet of things security: a position paper." *Digital Communications and Networks* 4.3 (2018): 149–160.
18. Dai, Hong-Ning, Zibin Zheng, and Yan Zhang. "Blockchain for internet of things: a survey." *IEEE Internet of Things Journal* 6, no. 5 (2019): 8076–8094.

19. Pan, Jianli, Jianyu Wang, Austin Hester, Ismail AlQerm, Yuanni Liu, and Ying Zhao. "EdgeChain: an edge-IoT framework and prototype based on blockchain and smart contracts." *IEEE Internet of Things Journal* 6, no. 3 (2018): 4719–4732.
20. Kshetri, Nir. "Can blockchain strengthen the internet of things?." *IT Professional* 19, no. 4 (2017): 68–72.
21. Christidis, Konstantinos, and Michael Devetsikiotis. "Blockchains and smart contracts for the internet of things." *IEEE Access* 4 (2016): 2292–2303.
22. Alotaibi, Bandar. "Utilizing Blockchain to overcome cyber security concerns in the Internet of Things: areview." *IEEE Sensors Journal* 19, no. 23 (2019): 10953–10971.
23. Reyna, Ana, Cristian Martín, Jaime Chen, Enrique Soler, and Manuel Díaz. "On Blockchain and its integration with IoT. Challenges and opportunities." *Future Generation Computer Systems* 88 (2018): 173–190.
24. Alaslani, Maha, Faisal Nawab, and Basem Shihada. "Blockchain in IoT systems: end-to-End delay evaluation." *IEEE Internet of Things Journal* 6, no. 5 (2019): 8332–8344.
25. Sahmim, Syrine, and Hamza Gharsellaoui. "Privacy and security in internet-based computing: cloud computing, internet of things, cloud of things—a review." *Procedia computer science* 112 (2017): 1516–1522.
26. Ali, Muhammad Salek, Massimo Vecchio, Miguel Pincheira, Koustabh Dolui, Fabio Antonelli, and Mubashir Husain Rehmani. "Applications of blockchains in the Internet of Things: acomprehensive survey." *IEEE Communications Surveys & Tutorials* 21, no. 2 (2018): 1676–1717.
27. Liang, Xueping, Juan Zhao, Sachin Shetty, and Danyi Li. "Towards data assurance and resilience in IoT using blockchain." In MILCOM 2017-2017 IEEE Military Communications Conference (MILCOM), pp. 261–266. IEEE, 2017.
28. Wu, Mingli, Kun Wang, Xiaoqin Cai, Song Guo, Minyi Guo, and Chunming Rong. "A comprehensive survey of blockchain: from theory to IoT applications and beyond." *IEEE Internet of Things Journal* 6, no. 5 (2019): 8114–8154.
29. Kamran, Muhammad, Hikmat Ullah Khan, Wasif Nisar, Muhammad Farooq, and Saeed-Ur Rehman. "Blockchain and Internet of Things: abibliometric study." *Computers & Electrical Engineering* 81 (2020): 106525.
30. Ferrag, Mohamed Amine, Makhlouf Derdour, Mithun Mukherjee, Abdelouahid Derhab, Leandros Maglaras, and Helge Janicke. "Blockchain technologies for the internet of things: research issues and challenges." *IEEE Internet of Things Journal* 6, no. 2 (2018): 2188–2204.
31. Sengupta, Jayasree, Sushmita Ruj, and Sipra DasBit. "A comprehensive survey on attacks, security issues and blockchain solutions for IoT and IIoT." *Journal of Network and Computer Applications* (2019): 102481.
32. Li, Shancang, Tao Qin, and Geyong Min. "Blockchain-based digital forensics investigation framework in the Internet of Things and social systems." *IEEE Transactions on Computational Social Systems* 6, no. 6 (2019): 1433–1441.

5 AI and IoT Integration

Gnanaprakasam Pandian,
Vivek Vinayagam, Brian Xu, and Mark Sue
Ordr.net
Santa Clara, California

CONTENTS

5.1 INTRODUCTION

As enterprises continue to deploy various types of devices, automating the data collection by device type and building the device behavior models can increase the efficiency for those focused on device visibility and security management. All these devices generate massive volumes of data, and each type of device uses unique protocols and varying types of traffic. In this chapter, we discuss how machine learning can help process volumes of disparate data in an efficient manner to identify the various device type and understand their behavior. This buildout of the device behavior models will simplify the job of network administrators to cope with the numerous and disparate types of devices and help them manage and secure the devices.

With the intention of device discovery and device behavior, a well-designed artificial intelligence (AI) and machine learning (ML) system ingests volumes of data, learns, and makes decisions. AI refers to a broad class of software, that is, capable of modifying its function based on prior experience. The steps listed in Table 5.1 is

TABLE 5.1

Machine Learning Pipeline

	Pipeline Stages
1	Data acquisition
2	Feature engineering
3	Data preparation
4	Model selection
5	Model training
6	Model testing
7	Parameter tuning
8	Model deployment
9	Model monitoring
10	Application development

a classic depiction of an ML pipeline, where we show how the acquired dataset goes through a transformation phase at each that culminates in the buildout of device models. Once these models are built and deployed, they can be used for making intelligent decisions in the field. Table 5.1 shows the stages in the pipeline, starting with training and completing with Inferencing:

5.2 MACHINE LEARNING FOR LEARNING MACHINES

Taking the general ML Pipeline as the baseline, we discuss how we can build an ML pipeline for device classification and device behavior. The main purpose is to understand the qualitative nature of connections and behaviors of these devices. In a dynamic network environment, an intelligent model learns what is normal and abnormal and combines the information with the classification and feedback data that further improves the overall network system integrity. Here we will also discuss how device clustering and classification is very important as this forms the building blocks for easy device management and anomaly detection.

At a high level, a well-designed AI system consumes packets from a sensor appliance, that is, connected in line with a distribution switch and gathers metadata information in a process known as feature extraction. Using an unsupervised-learning algorithm (without human intervention), the AI system clusters these devices into groups of similar behavior and flow. Once these devices are grouped into clusters, an artificial neural network (ANN) based supervised model is used to get trained the inter/intra-cohesiveness of devices among the groups, later the model is used to classify/profile the devices in inventory on the fly.

An ANN is a machine learning algorithm that is, modeled from the human brain and what is currently known about how the brain functions. The brain consists of billions of neurons interconnected to form neural networks responsible for carrying out cognitive tasks such as identifying human faces, grouping entities into categories, solving problems, and making decisions.

There will be instances when a new device is found that does not match an existing ANN model as new devices are being constantly created and connected. In this event, the classification process proceeds to the nearest hierarchical profile, that it could match with a set confidence score. These new devices are then exported to a learning cloud engine, which acts as a profile exchange where multiple customer device profiles are kept in a common repository.

5.3 HOW IS AI USED FOR DEVICE PROTECTION

There are too many types of devices and subcategories when one considers the amount of IoT connections in the enterprise work environment. As estimated by IDC [1], this number is expected to grow by up to 28% per year through 2025. AI in an environment of heterogeneity and increasing volume can provide quick insights from the data that is ingested. ML and AI technologies bring the ability to rapidly automate the identification of patterns to detect anomalies in the data that is ingested.

The technology goes beyond the clarification of known devices and categorizing new devices. AI can also be implemented to analyze the data that is generated, the traffic between types of devices, and the behavior of certain device classes. Compared to traditional methods, ML overall can help ingest a lot of information and offer operational predictions at multiple higher rates.

An architecture of AI and IoT integration is illustrated in Figure 5.1 that consists of four layers: third-party tools, data lake, AI–ML processing, and security layers.

In this architecture, network and cybersecurity tools (e.g., Cisco, Fortinet, etc.), and third-party threat intelligence (TI) feeds are fed into data lakes that provide storage and archival services, device adaptation services, and feedback commands to the network and security tools. Importantly, the data lakes send the processed datasets to the AI—ML processing layer, that is, integrated with network security tools and technologies.

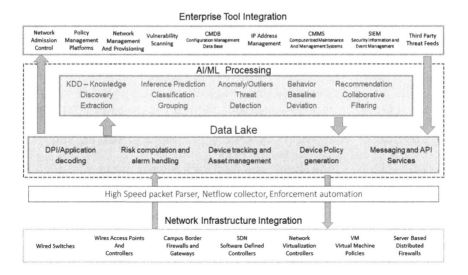

FIGURE 5.1   The architecture of AI and IoT integration.

The AI–ML processing is the core layer that provides the key capabilities for IoT device security services such as (1) KDD—Knowledge discovery and extraction; (2) Inference prediction and classification grouping; (3) Threat detection of anomalies and outliers, (4) Behavior baseline and deviation, and (5) Recommendation, collaborative filtering.

5.4 ML PIPELINE

We discuss some of the important stages in the ML Pipeline in this section.

5.4.1 DATA ACQUISITION

Device classification occurs when there is an understating of applications and protocols moving within a network environment. In Table 5.2, we see how a knowledge graph is constructed based on the list of data sets acquired by various means. An extensive library is analyzed and created, and the classification occurs at a global level. When it comes to the granular details, an extensive list of communications and management standards is taken into consideration. For example, in imaging, observing the digital imaging and communications in medicine (DICOM) [2] standard for medical imaging information and related data is used. There are several other protocols used in a hospital or medical environment, which we highlight below:

DICOM: DICOM [3] is often used for storing and transmitting medical images enabling the integration of medical imaging devices. These devices can include scanners, servers, imaging systems, and picture arching and communication systems (PACS) from various manufacturers. DICOM is widely adopted by hospitals providing many files that can be analyzed and observed.

TABLE 5.2
IoT Device Datasets and Acquisition

Aspect	Components
Device probes	SNMP, UPnP, MDNS, WinRm, Windows PowerShell, OS, AV SW, Patches
Network packet data	Routers/Switches/Wireless, NETBIOS, logins, file transfer, HTTPS, P2P user agents, certs
IoT protocol decodes	DICOM, POCTX, Baxter, Patient Monitoring, GE Muse, BACNET, Modbus, Profinet
Control plane data	3000+ Protocols/Apps, ARP, DHCP, MDNS, LLDP/CDP, Traps and Logs
Utilization and performance	Number of studies/scans, battery level, traffic volume, session count, TCP retries, connection resets
Opt-in scan for non-mission critical devices	Open Ports, Software Vulnerabilities, Outdated Software, Weak Passwords
User information	Active Directory, Radius, LDAP, local users
Net device info	Netflow/SFLOW/IPFIX, Cisco ISE, Aruba Clearpass, ForeScout, WIFI Controllers/APs, Firewalls/PAN/ASA/Fortient

---

HL7: Another standard for the exchange of electronic health care data is Health Level-7 (HL7) [4], which was created by the non-profit organization Health Level Seven International. HL7 provides guidelines and standards to help software and healthcare providers to store and uniformly move data. Applications can use the data without special conversion software.

Interestingly, the HL7 standard is sometimes referred to as the "no-standard standard". Often times, different healthcare organizations implement the HL7 framework in unique ways due to the absence of a standard clinical or business process for interacting with clinical data, patient records, or relevant personnel data. Regardless, HL7 aims to make sharing data easier for medical administrators, and we can readily see and analyze the flow of HL7 traffic in these networks.

RTSP: For digital cameras, a specific protocol RTSP (Real-Time Streaming Protocol is used. RTSP is an application-level protocol for control of the delivery of data with real-time properties. RTSP provides a framework to enable controlled, on-demand delivery of real-time data, such as audio and video. In RTSP, the sources of data include live feeds and archived clips. The protocol is intended to control multiple data delivery sessions and to provide a means for choosing delivery channels such as UDP, multicast UDP, and TCP.

IPP: Printing applications often use the Internet Printing Protocol (IPP), which is a protocol for communications between devices such as computers, laptops, mobile devices, and printers or print servers. The protocol allows clients to submit one or more print jobs to the printer or print server. Tasks can include querying the status of the printer in the network, getting the status of a print job, or the ability to cancel individual print jobs from a queue. IPP also supports access control, authentication, and encryption, making it a more secure protocol than older protocols. IPP uses TCP with port 631 as it's a known port.

BACnet: For HVAC use, the BACnet (Building Automation and Control Networks) protocol was developed under the guidance of the American Society of Heating, Refrigerating and Air-Conditioning Engineers (ASHRAE). BACnet is an American national standard, a European standard, a national standard in more than 30 countries, and an ISO global standard. BACnet IP UDP is the UDP Port number used by BACnet communications over IP.

There are numerous protocols moving around a network—all of which provide ample data for learning engines. A learning engine can be constantly updated and supervised and refined in the broad view based upon all of the information that is observed. Aside from the protocols, the AI system sees requests and traffic patterns that are also applied to its learning engine. If a device, for example, reaches out to Amazon.com and matches certain other criteria in its library, the classification model may determine that this device is, in fact, an Alexa device and search for other supporting or disproving attributes.

Deep Packet Inspection [5]: Previous forms of packet filtering only looked at header information, which, to use an analogy, is the equivalent of reading addresses printed on the outside of an envelope. This was due partly to the limitations of technology. Until recently, firewalls did not have the processing power necessary to perform deeper inspections on large volumes of traffic in real-time. Technological advancements have enabled DPI to perform more advanced inspections that are more like opening an envelope and reading its contents.

Also, using DPI to examine the contents of packets passing through a given checkpoint in a network, does not introduce any latency into the network. The system can make quick decisions on the traffic flow based on what the packet contains. Traditionally, various methods of packet filtering looked at the packet header information and just the rudimentary data.

DPI takes the technology further by enabling the learning system to look inside and see the contents of information traveling within the network. DPI is also used for the detection and interception of viruses and other forms of malicious traffic to help keep an enterprise network safe. DPI overall is used in network management to streamline the flow of network traffic as certain traffic may have priority over basic web browsing, for example.

5.4.2 FEATURE EXTRACTION

Learning models become more intelligent and improve over time when the amount of data increases, and the data is continuously refined as it is ingested. This is the best accomplished by analyzing and observing all the various protocols that are crisscrossing the network internally and even the traffic that leaves the enterprise (see Table 5.3).

Deep Packet Inspection is used to look into the packet. This data is combined with a growing library of asset management systems. Behavior models are constantly

TABLE 5.3

Context of the Device at Various Layers of the Software Stack

Layer	Attributes
Application	Where: Location, geographic coordinates
	Network: Topology
	Utilization: Peak, low
	Detect: Keepalive
Transaction	Sequence: Events
	Trigger: Signature, anomaly
	Detection: Performance issues
	Attempts: Login Failures
Flows	To whom: Peer
	From whom: Device
	Why: Transaction
	How: Protocol used
Device	Who: Device/User
	What: Method
	When: Timestamp
	How: Much, often, long

created and collected, and supervised learning takes place whereby the data is connected to a cloud central model, which then feeds the previous device profiling engine in a loop.

5.4.3 GROUPING—CLUSTERING ALGORITHMS USING UNSUPERVISED LEARNING

Clustering is an efficient tool when it comes to grouping similar IoT devices. The algorithm can be set to run continuously and, as the packets continue to flow within the network environment, the process keeps learning and grouping as required. The limitation, of course, is that there may be an instance when the clustering may converge on the identity of what a device is, and the algorithm may not match a global profile. At that point, a sub-segmenting classification process is undertaken when it comes to fine-tuning the model.

Unsupervised learning is when the machine is trained using information, that is, neither classified nor labeled and the algorithm can act on the information without guidance. The AI system is tasked with grouping unsorted information according to similarities, patterns, groups, and differences without any previous training of data. No guidance is provided, no teaching is provided, and the machine is restricted to find structure in the unlabeled, unstructured data.

As illustrated in Figure 5.2, unsupervised learning is typically classified into two categories of algorithms, namely clustering and association. In clustering, devices are placed in groups of similar behavior and flow. The problem to solve is when a user wants to discover the inherent grouping in the data, such as grouping devices by the type of behavior within a network. Association, in contrast, is learning that makes inferences based upon rules that describe large portions of the data. An example of this is Netflix's environment, where people that view movie X also tend to view movie Y.

5.4.3.1 Centroid-Based Clustering

In Centroid based clustering, the data is organized into non-hierarchical clusters. K-means is the most popular centroid based clustering algorithm and the model is efficient when large amounts of data are considered. The issue with centroid based clustering, however, is that the algorithm is sensitive to initial conditions and outliers.

5.4.3.2 The Divisive Method of Hierarchical Clustering

A flat clustering subroutine process is used to split each cluster until it has data having its own singleton cluster when it comes to the Divisive Method of Hierarchical Clustering. This top-down method works well as it is not required to specify ahead of time the number of clusters. The model does become efficient when one generates a routine at the start of a complete hierarchy all the way down to individual data leaves.

Divisive algorithms can be more accurate than the Agglomerative clustering mentioned below since Agglomerative clustering makes decisions by considering the local patterns or neighbor points without initially considering the global distribution of data. One needs to know the big picture to structure the algorithm in the beginning. Subsequently, early decisions cannot be undone, whereas divisive clustering

By using unsupervised ML techniques to
enrich data with learned attributes,
observed devices will cluster around
known profiles.

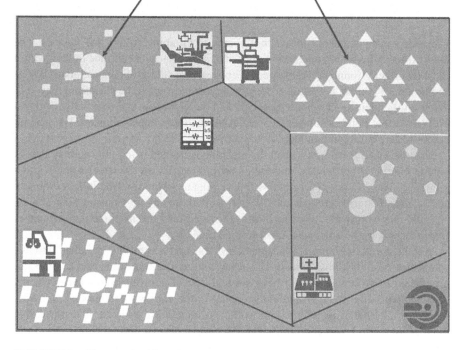

FIGURE 5.2   Unsupervised learning clustering and association.

takes into consideration the global distribution of data when making the top-level partitioning decision.

5.4.4   IoT Device Classification

"It is not who I am underneath, but what I do that defines me"—BATMAN.

Considering the volume growth of the number of devices such as medical, industrial, or basic IoT, "unsupervised" machine learning can be very efficient in grouping similar devices. In a networking environment where a sensor is placed within a SPAN tap to observe the traffic, the clustering model provides the ability to group devices into various "buckets" before the classification process starts.

With IoT devices, a methodology to classify and organize the data into relevant categories or groups can make it helpful considering the plethora of details that can be attributable to each device. Classification of data and placing the information into groups and subgroups makes it easier for access and analysis. This process can sometimes involve tags or labels and a library to understand the organization of these groups.

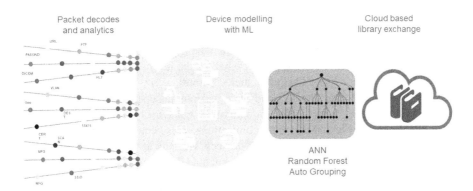

Packet decodes and analytics

Device modelling with ML

Cloud based library exchange

ANN
Random Forest
Auto Grouping

FIGURE 5.3   IoT Device classification models.

Data classification often involves a multitude of tags and labels that define the type of data, its level of confidentiality, and integrity. Availability may also be taken into consideration in the data classification process. Data's level of sensitivity can further be classified based upon varying levels of importance or confidentiality, which then correlates to the security measures put in place to protect each classification level.

From a mathematical modeling point of view, data classification is the process of predicting the class of given stated data points. Classes within data classification are often stated as targets or labels or even categories for that matter. In the case of IoT, the classes can be the type of machinery, the function of the device, or even the access level of these devices. Once enough data is invested, classification modeling is used to predict the next series of data that involves the task of mapping input functions to discrete output varies.

In basic instances, the process can be a binary classification if something is either one object or not that specific object. For example, a visual categorization model for a cat will use data sets of a cat and non-cats as the training data. The classifier will use this training data to understand how given input varies to the class of data. When the visual classifier is trained accurately and with granular detail, it can be used to detect an animal such as a dog, which in this case, is a non-cat but still an animal. Classification models can be binary or have multiple classes, as depicted in Figure 5.3.

As detailed in Table 5.4, supervised learning [6] vs. our prior discussion of unsupervised learning is a subset of ML whereby algorithms are created to learn by example; hence the term "supervised". The training data typically consists of inputs tied to the correct outputs in a supervised learning model to give the program assistance.

TABLE 5.4
Supervised Learning Categories

Type	Method
Continuous target variable	Regression
Categorical target variable	Classification

In the training process, the supervised learning algorithm will ingest new previously unseen inputs and then determine under which label the new inputs should be classified based upon the prior training data. The goal of the supervised learning model is then to predict with the correct label for the newly presented input data (a function of the input data yields the predicted output).

Beyond understanding devices and using various unsupervised and supervised learning models, the system looks in detail at the behavior of devices as we understand the flow of traffic and their interconnections. This behavior modeling also utilizes similar learning models to analyze devices. We look at what the device should do and what the outlier behavior might be, such as deviation within category, subcategory, profile, or group. Learning the appropriate behaviors between and among devices helps us understand what is appropriate and when the behavior is deemed inappropriate.

When a new device appears in the network, the learning engine will generate its behavioral model and compare it with known behavioral models for device classification and authentication. Since the behavior has many attributes, we are, in fact, performing multi-factor authentication as well as continuous authentication. The system works closely with well-known sources to identify and see deep into signatures for anything that might provide clues of ill intent. The learning system relies on signatures to identify what is "known" malware. Classification and profiling capabilities and features include (1) Packet decodes and analytics, (2) Device modeling with ML, (3) Auto grouping, and (4) Cloud-based library exchange.

5.4.4.1 Classification Technique—Random Forest

One can visualize a Random Forest supervised learning algorithm as being similar to that of a spanning tree in a network. The forest is essentially a large collection of decision trees that are merged together to help to increase the accuracy of predictions. The advantage of a Random Forest is that it can be used for both classification and regression problems.

In a classification sense, The Random Forest technique has similar parameters as a decision tree or a bagging classifier. In this situation, there will be no requirement to combine a decision tree with a bagging classifier. One can use the classifier-class of the random forest. (A Bagging classifier is an ensemble meta-estimator that fits base classifiers—each on random subsets of the original dataset and then aggregates their individual predictions {either by voting or by averaging} to form a final prediction.).

Random Forests can add additional randomness or unpredictability to a model while growing the trees. Instead, for example, of searching for the most important feature while splitting a node, it searches for the best feature among a random subset of features. This results in a wide diversity that can result in an improved model. Therefore, in a random forest, only a random subset of the features is taken into consideration by the algorithm for splitting a node.

5.4.4.2 Classification Technique: Various Neural Networks

There are several AI Neural Network models, such as the recurrent neural networks (RNN) and convolutional neural networks (CNN). RNN takes the basic model of a neural network and modifies it to consider information from the previously entered data and the currently entered data to perform a task. This is a model that works well, for example, in natural language, tasks as a word meaning in a full sentence is often highly dependent on the previous words in that sentence. RNNs perform this algorithm by keeping an "internal state", that is, used to save the context of the data being fed into the network and this state-issued to predict the outputs for the subsequent inputs.

CNN is another deep learning technique, that is, gained more popularity in deep packet inspection. CNN classifies input data based on the input features and then assigns the data to a category. Deep packet inspection by CNN starts with a 1D CNN layer with filters of size 3. The filters convolve over the input data with a stride of 2. The output of this layer is called the activation map. After the activation map, max pooling is done, followed by dropout. Eventually, the next layer is then fed into a series of fully connected layers. Finally, a SoftMax classifier is used to performs the classification task predicting the device types.

5.5 ANOMALY DETECTION

Anomaly detection refers to data flow, observations, or events that do not conform to the expected pattern of a given device or behavior. Figure 5.4 shows anomalies detection steps, and Table 5.5 shows different types of anomalies in IoT devices.

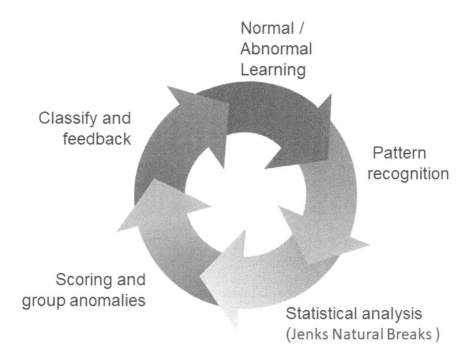

FIGURE 5.4 Anomaly detection.

TABLE 5.5

Different Anomalies of IoT Devices

Risk Level	Risk score	Description/Example Events
Conforming to baseline	Blue	Certified HW, predictable behavior, runs approved SW Corporate-sanctioned devices and inventory tracked.
Known good state	Green	No observed deviation from baseline and compares will with peer groups and past trends. No new external or internal communications, that is, suspicious.
Slight increased risk	Yellow	Known good state but some flows and behaviors are suspicious. Internal cross traffic and/or C&C attempt.
Needs to take action	Orange	Surely suspicious. Weak posture or cryptography. Alerts to SIEM systems, flawed traffic patterns.
Needs immediate attention	Red	High severity alerts, SMS alarms, trouble ticket raised. Isolation and segmentation required, data exfiltration.

These colors are described as follows:

Blue—*More than N devices exhibit this behavior.*

- In most cases, enterprises will have more than N devices in a profile, if they are medical or other mission-critical devices.
- A clustering algorithm (totally unsupervised ML) creates these clusters within hours of traffic flowing, and hence the same size increases to N quickly.
- In general, this is within the local enterprise-level behavioral profiles.
- Here the match is to the exact port number, exact IP address, exact gateway, etc.
- Golden profile ports/protocols are not relevant here as we have a decent size sample, within the context of the enterprise. The idea here is that a hacker compromising a specific device of a specific profile is unlikely to compromise the same type of device in a different part of the organization. Usually, recon attempts to exploit vulnerabilities of devices in their network vicinity, and the malware does not know what type of devices it is exploiting to serve as a weapon.

Green—*For a given destination/port/protocol—less than N devices exhibit this behavior and found a match in global profile.*

- This usually applies to many consumer-class devices and facility-type IoT devices.
- Compare against global profiles to determine if any of these applications/ protocols/ports are normal, as exhibited by the collective list of all customers.
- If other devices of similar type exhibit this behavior—risk rating stays Green.

Yellow—*less than N devices exhibit this behavior and*
no match to the global flow-level profile.

- If not seen—risk rating goes up to yellow.
- The only caveat here is that a global profile would have generic information like protocol, gateway, and so on.
- N is configured to be five but is adjustable. Anomaly detection with respect to global profiles above and beyond baseline deviations—Exceptions for flows within the enterprise.

Orange

- Traffic flow from one to an excessive number of targets—spreading malware.
- Traffic flow from many compromised machines to a single target—DDOS attack.
- DNS morphing—we use our own algorithm to detect DNS reach-outs and DNS-based tunneling.
- Anomaly detection for flows going in/out of the enterprise—behavior baselining.
- The global profile URLs are used heavily here. It is very rare that a specific enterprise needs to reach out to a specific site, that is, not exhibited in other customer sites.
- The idea of N devices exhibiting the same behavior does not apply here. Any device type with any URL is subject to the following conditions to trigger the anomaly. Also, just because of all the devices within a device type all reach to the same URL, it does not mean the URL is OK. We usually find that most devices within a profile reach out to the same URL, for example, when a firmware update occurs.

Red—*Any traffic subjected to intrusion detection rules*
with signatures sourced from well-known vendors.

- This serves to immediately detect malware like WannaCry, DoublePulsar, Trojans, DDOS, etc., — risk rating immediately raises to RED.
- In GUI, this is shown in the threat cards with a risk rating.

Quantified scoring of anomaly detection

- Red—If the URL reputation score is tied to a risk rating score (>70 = yellow, >80 = Orange; >90 = Red)
- Red—Based on data rate, that is, normal for this profile—If the data rates exceed the usual rate seen in any customer site—the risk rating goes red.

- Read—the machine is talking to Command & Control site—as given by the URL reputation score—then the risk rating goes red.
- Orange—URL reputation is >80.
- Yellow – URL reputation >70.
- N today is configured to be 5, which is adjustable.

There are numerous anti-malware solution providers that identify objects and add new signatures to their known databases 1. These repositories grow each day and hold data on hundreds of millions of signatures that identify and classify malicious objects. Signature protection against malware works, and it is relatively easy to use. It is a tried and true method of catching the millions of older but still persistent threats that are roaming out there.

5.5.1 Not Just Matching Signatures but Going Beyond

The problem in this sophisticated age of cyber-attacks is that some versions of code may not always be recognized by the traditional approach of mapping signatures. New versions of dangerous code can appear that are not readily recognized by traditional signature-based technologies. A study by Cisco found that 95% of malware files analyzed were not even 24 hours old. Worse yet, sometimes signatures can morph and hide. With some code permutations: a change here, a register renamed there, or code shrunk or expanded, and malware can avoid the traditional signature detection.

Because a well-designed learning model is trained, it knows the appropriate actions of a device and what that device should be doing—be it sending images and video streams to a video server at the data center or communicating with a print server that has been identified in a network. If there is any deviation from known behavior, it will be seen right away and shut down immediately in a proactive manner.

It is not just the changes from the daily routine that can be observed a well-designed engine can monitor how each device acts in a normal setting relative to its peer group. Time series are factored into what can be seen if a camera's behavior, for example, is deviating or is different from what its behavior was in the prior weeks.

If there is some previously unobserved communication between a camera and remote suspicious site, a proactive system can prevent this since it is smart enough to know that a camera never had this session in the past. Even if there is a request going into the camera from an external site, it will sound the alarm. Is there an attempt to extract a video without permission with data leaving the network? This is a behavior violation. If it's a behavior violation that has been learned via the algorithm and is now a rule, the system will prevent it immediately. A thermostat talking to the finance department should not happen, for example, as it is a behavior that is inappropriate. Some traffic flow trying to disable security controls or install rootkits is yet another example that the model will flag.

TABLE 5.6

Policy Generation

VMware/NSX	DC server rules based on device grouping
Firewall and AD	Zone policy—restrict flows towards internet, Pushing policies to AD for IoT
ISE/CPPM/ForeScout	Auth policy—CoA, VLAN isolation, TrustSec, SGT tags, ACLs, group-based rules, DNS filters
Core/Routing	VRF, Routing, Private VLAN policy
WLAN	Wireless Devices—Flow whitelist, MAC blacklist, Location-based policy
Distribution	Subnet/Edge overflow policies, Peer group-based authentication
Access	VLAN isolation, flow whitelisting, Port security, Port shutdown

5.5.2 Continuous Learning Driven By AI

Threats are dynamic and constantly evolving. Having a system that understands signatures can help in a hyperconnected environment, but to have real proactive protection, a system must be aware of contextually and have the insight to understand behaviors. Diving deeper into historical patterns can also help capture baseline deviations that might fly under the radar.

The system is constantly expanding its behavior library, understanding what is normal and what is out of character for each device type. It should complement current tools and work with what users already have. The system helps you quickly identify all the devices by make, model, and number, and all the traffic that it generates.

When devices are understood and correctly grouped, and it is important to spend much more time with understanding and learning about the behavior of the device and not just the device itself. The learning and monitoring of the device behavior can uniquely identify the device and perform authentication of the device. ML models are utilized to generate a more complete, relevant, and up-to-date device profile and, subsequently, a baseline is generated for anomaly detection. Security policies (see Table 5.6) are also generated for networks with protocols, security software, and tools to detect anomalies and threats.

5.6 CONCLUSION

AI and ML have a critical role to play in the task of defending the networks of health care institutions from security threats. The process of profiling devices, capturing network information, and identifying threats are all necessary steps that require AI/ML to cope with modern network conditions and continuously evolving malware. Given that the medical networking infrastructure is not only business-critical but vital to saving lives; it has become urgent to install defensive measure that is responsive and effective. This chapter showed how this can be done.

REFERENCES

1. https://www.idc.com/getdoc.jsp?containerId=prUS45213219
2. https://en.wikipedia.org/wiki/Supervised_learning
3. https://en.wikipedia.org/wiki/Unsupervised_learning
4. https://en.wikipedia.org/wiki/DICOM
5. https://en.wikipedia.org/wiki/Health_Level_7
6. https://en.wikipedia.org/wiki/Deep_packet_inspection

6 Security Challenges of IoT and Medical Devices in Healthcare

Gnanaprakasam Pandian, Vivek Vinayagam,
Brian Xu, and Mark Sue
Ordr.net
Santa Clara, California

CONTENTS

6.1 INTRODUCTION

Growing preference for real-time data monitoring and network-connected medical equipment among healthcare institutions is an established and accelerating trend. Market research firms IDC [1] and Grand View Research [2] are projecting the IoT healthcare market to reach USD $534.3B, growing at a compound annual growth rate (CAGR) of 19.9%. At the same time, the general IoT market is projected to grow at a CAGR of 28.7%, ultimately generating nearly 80 zettabytes (ZB) of data per year by 2025.

Hospital systems are becoming massive in scale with correspondingly large networks. For example, Taiwan's Chang Gung Memorial Hospital with 10,000 beds. Each bed could have up to 15 IP-connected devices providing continuous monitoring. As a result, this typical modern healthcare environment must support tens of thousands of IoT devices that are being connected to the network. This is a trend that will continue with digitization. Not only will more devices be connected to the network, but many components will require access to application servers and database storage. In addition, some of these devices will require access to internet destinations such as hosted services, including advanced analytics and image enhancement services.

Ransomware attacks on healthcare institutions are increasing. The WannaCry variant of ransomware is the one, that is, the most often used over the last few years. The most important points of this trend are as follows:

- The year 2019 was specifically notable in that:
  - The healthcare sector has taken some massive hits when it comes to cybersecurity, big hacks, and third-party vendor breaches.
  - Ransomware also saw a resurgence; more than 750+ incidents have been recorded.
  - WannaCry across the sector is quite prevalent and widespread, as evidenced by the number of published disclosures healthcare institutions are making.
  - Patient records for hundreds of healthcare providers leaked, and press releases on these disclosures continue to ramp up.
  - Patient care disruptions due to the targeted, sophisticated nature of the attacks put tremendous strain both on the providers and on their IT vendors.
- Disruption to health and safety is paramount as the incidents were not simply expensive inconveniences. The actual damage observed includes:
  - Emergency patients to be redirected to other hospitals when critical imaging machines go down.
  - Medical records were inaccessible and, in some cases, permanently lost. Putting front end patient admissions office staff in jeopardy.
  - Surgical procedures were canceled, tests were postponed, and admissions halted due to the unavailability of systems that deal with both hospital management systems as well as patient healthcare systems.

These trends are due both to the proliferation of security vulnerabilities and the increased sophistication in attacks, making it essential to think about prevention and detection. Healthcare organizations are mission-critical and cannot take this lightly.

6.2 SPECIFIC SECURITY CHALLENGES IN HEALTHCARE

Networks and networked systems that serve healthcare institutions are subject to security considerations that are in addition to business networks in general. These considerations include:

- A large and continuously changing number of healthcare related devices that are network connected.

TABLE 6.1
Differences between Business and Healthcare Devices

Behavior	Medical Device	Business Device
Logging on to network	Registered as a medical device	Registered as a business device
Serial number	Matches medical device manufacturer	Matches business device manufacturer
User credentials	User is an authorized medical technician.	User is an authorized business user.
Device control/ownership	Medical technology department	IT department
OS software and version	Consistent with the requirements of OEM manufacturer.	As required by business applications and IT security concerns.
Scans	As per manufacturer specifications.	May change when users install applications.
IT security agent	Provided by OEM manufacturer only.	As per IT department policy.

- Medical devices store and transmit personal health information (PIH), which is a frequent target of malware attacks.
- Medical devices require certification and calibration on a mandated schedule.
- Connected medical devices can be highly mobile within the campus perimeter.
- Some medical devices (specifically medical workstations) are subject to misuse. For example, users of workstations can use browsers installed on these systems for personal use.
- Imaging devices will routinely transmit large files through the network.
- Devices in medical institutions can represent a direct health threat to patients if they are compromised.

Simultaneously, the steady migration of IT to the cloud has increased the interactions outside the traditional corporate network, making security even more of a focus in the healthcare industry, such as hospitals. Malware exploits mean not only a loss of operational capability but also a large potential liability if data is compromised or stolen.

Table 6.1 presents some of the differences issues that are present in traditional IT system like a laptop or a personal computer (PC) as opposed to networked medical devices like a medical resonance image scanner (MRIS) machine.

6.3 CLASSIFICATION OF IoT AND MEDICAL DEVICES

6.3.1 CHALLENGE 1: CLASSIFICATION OF IoT AND MEDICAL DEVICES IN HEALTHCARE

The expansion of the use of medical IoT devices that are being connected at hospitals includes electrocardiograms (EKGs, or ECGs), blood pumps, glucose meters, and multiple generations of legacy OT (Operational Technology) devices such as CT (Computed tomography) scanners, x-ray machines, ventilators, etc. Because of the growth in the number of devices and the distributed nature of modern networks, it is becoming increasingly difficult to manage IT, OT, and IoT (Internet of Things) separately.

Typical device attributes collected in the classification system include any and all of the following:

- Manufacturer/make/model/modality.
- Typical OS type/firmware version.
- Public advisories such as Food and Drug Administration (FDA) recall. Other government authorities may also issue recalls.
- Manufacturer recall.
- Private (commercial) recalls by manufacturers
- National vulnerability database and common vulnerability exposure (NVD/ CVE) vulnerability information.
- Group, category, device profile—classification hierarchy.
- Typical external URL/web sites accessed.
- Default applications and network protocols for device type based on function—DICOM for imaging, HL7 encoding for all medical, RTSP for cameras, IPP for printers, BACnet for HVAC, etc.,
- Typical ports/protocols for this profile.
- Typical applications and server gateway types.
- Typical user profile/admin privileges—how many users, etc.,
- Set of mean statistics of traffic flow—for example, to detect data exfiltration.

6.4 CURRENT SECURITY CHALLENGES OF IoT AND MEDICAL DEVICES

Traditionally perimeter firewalls were erected to protect an enterprise network. As networks have evolved, protecting a border that has become amorphous, that is, merely physical has proven to be difficult. The protection of devices by implanting agent software was a temporary next step but limited in efficacy, considering that many devices do not accommodate an agent. These new classes of IoT and medical devices are not designed to work with third-party embedded security agents and cannot be patched and upgraded by the IT department due to FDA certifications or other regulatory reasons.

Another problem associated with IoT devices is that they are typically built with OEM versions of operating systems (Windows, Linux, or other) that do not necessarily receive the same security patches and upgrades as the products on which they are based. Documented security exploits may be left unpatched for longer periods of time. When patches and upgrades are made available, they are usually on a schedule, that is, specific to each device make and model.

Business continuity is a top requirement as the operation cannot afford any downtime. The current practice of "Authenticate the user and trust the device" is no longer adequate. Just because a device has the right credential to enter a healthcare network does not mean it has the authority to access whatever medical systems it wants at will. The challenge is not only the sheer number of devices but the variety of devices and dynamic environments with the astronomical growth of their deployment.

6.4.1 CHALLENGE 2: SECURITY CHALLENGES OF IoT AND MEDICAL DEVICES IN HEALTHCARE

The primary security challenge of IoT Devices is accurately assessing the risk in terms of network security. Device risks are defined by the formula:

$$\text{Risk} = f(\text{assets, threats, vulnerabilities}) — \text{controls}$$

where

Assets: An Asset is a data or a device. Assets are assigned a level of criticality from low to high. For medical devices, the default criticality is high.

Threats: Threats will include those detected by IDS (intrusion detection systems), including Malware, Trojan, Recons, or OS Exploits, Bad/Phishing URL Access, C&C (command and control) Communication, Data Exfiltration, Behavior Anomaly, Weak TLS (Transport Layer Security), and Cipher usage among others.

Vulnerabilities: Vulnerabilities include older operating systems, AV (anti-virus) not up to date, ICS-CERT (Industrial Control Systems—Cyber Emergency Response Team) advisories, FDA recall, MDS2 (Medical Device Security), Open Ports Detection, Weak Passwords, Software Vulnerabilities (through mostly using network-based passive scan techniques), and exploitable weaknesses in the design of the device.

Controls: Controls include OS/AV update, Blacklist MAC (Media Access Control), shut switch port, quarantine VLAN (virtual LAN), Flow Whitelisting/blacklisting, Automatic Access control, Policy Push to Switch/Firewall/ISE (Identity Services Engine), SIEM (Security Information and Event Management) and CMS (content management system) notification, Alarm workflow, and techniques used for micro-segmentation.

For healthcare, the Association for the Advancement of Medical Instrumentation (AAMI) adopted the risk score mechanism described above as a standard with further refinements.

6.4.2 RISK MANAGEMENT STARTS WITH VISIBILITY AND INVENTORY COLLECTION

Before getting into risk management, a survey of the healthcare institution and its networked operation is needed to clearly understand the following:

- What do I have? Identify, classify track, and locate inventory.
- What are they doing? Watch communications and spot anomalies.

TABLE 6.2
MRI Machine Risk Management Lifecycle

	Stage
1	Profile devices
2	Assess vulnerability
3	Detect threats
4	Regulate applications
5	Enforce policies
6	Utilize efficiently

- Are they vulnerable? Assess exposed weaknesses and risks.
- How do I protect them? Restrict transactions and segment.
- Are we using them efficiently? Use business intelligence to track usage.

The risk management life cycle, using a typical MRI (magnetic resonance imaging) is listed in Table 6.2.

Device owners need to know the risk assessment details for their devices by recognizing the following risk-level conditions as listed in Table 6.3.

Critical risk items need immediate attention. Actions taken by the system include high-severity alerts, alarm messages sent via SMS (short message service), and initiating trouble tickets. Other required actions taken will include isolation, segmentation, and data exfiltration.

Medium/low slight increased risks (score: 0.1–6.9) require on-going monitoring. To help with risk mitigation, the security monitoring system must determine the following: open ports, weak passwords, weak TLS (transport layer security),

TABLE 6.3
Risk-Level Conditions

Condition	Description
Conformance to baseline	The hardware is certified and shows predictable behavior. Only vendor-approved software is installed, and the device is tracked by the inventory control system.
Known good state	No observed deviation from the baseline, and the behavior is like peer devices and exhibits the same level of interactions as it has exhibited by the last few weeks. No suspicious internal or external communication is detected.
Increased risk (low to medium)	Known good state, but some flows and behaviors are suspicious. Internal cross-traffic and/or transactions with command and control web sites communication that could potentially lead to a ransomware infection.
High risk	Weak posture or cryptography. Alerts to SIEM systems constantly sent, flawed traffic patterns, etc., that require immediate action.

and cipher usage, expired certificates. Presence of Known vulnerabilities, ICS-Cert advisories, FDA recall notices CVSS (common vulnerability scoring system). OS exploit/software vulnerabilities detected (OpenVAS CVSS).

The best practices that are used for assessing and fixing vulnerabilities by using patches, hotfixes, and anti-virus software updates. A common language among all the departments (Line of Business, IT/Networking, Security/Privacy and Clinical Engineering/Quality, and Legal and Compliance) to monitor and manage medical devices is needed.

The best practices that are used to assess vulnerabilities include:

1. Check the OS version to determine any published vulnerabilities in the national vulnerability database of FDA recalls.
2. Check for weak/default passwords and force updates.
3. Run any available vulnerability scans as part of the device onboarding process and/or periodic validation. Close any unneeded open TCP/UDP ports if possible.
4. Shut down unwanted services on the machine with priority to services that run with administrative access.
5. Understand its implication on active directory users and domain joins.
6. Patch (OS/AV) vulnerable MRI devices; identify older OS such as Windows XP and Windows 7 and upgrade and check for hotfixes.

A practical case is shown in Table 6.4 in how to assess the vulnerabilities by using a wide variety of parameters, context, hotfixes, and AV updates.

On top of the security risk assessment, one must consider the clinical risk of the medical devices when designing a system.

TABLE 6.4
Assess Vulnerabilities Using the Following Parameters

Contextual Area	Details of Parameters
HW inventory	CPU, memory, disk, BIOS versions, encryption levels.
SW inventory	OS version, patch level, AV software installed, security agent software, other software running on the system.
User information	Local users configured, active directory-driven users, accessing the system.
Public database	FDA recalls for this device.
Manufacturer data base	Manufacturer recalls for this device.
Scanning sources	Vulnerability info gathered from vulnerability canners like Tenable and Rapid7.
CVE data	OS version correlating to well-known published vulnerability in the national database.
Criticality	Medical and industry mission-critical devices are rated as high.
Alarm profile	Alarms that may be generated by this device.
Risk assessment	Risk score of the device.
Clinic assessment	Clinical risk assessment of this device.

- PHI Risk (3 categories)—Different organizations have identified key fields that flag PHI (about 20 or so) on each model of the device.
  - Accessibility of the equipment—Is it bolted down, such as a CT scanner, somewhat mobile such as an ultrasound, or totally portable such as a laptop?
  - How many PHI fields are on the device 1–3, 4–8, or 9 or more; any DICOM device has 1750 fields.
  - How many records can the device retain 1–50, 51–499, over 500. If over 500 and the device is lost, stolen, or hacked, these devices are immediately reportable and carry major fines.
- Clinical risk (5 categories)—Defined "if device stops working or fails can cause death or impairment".
  - Life support is the highest including, defibrillators, ventilators, anesthesia units.
  - High risk devices such as laboratory analyzers, imaging equipment, IV pumps.
  - Medium risk devices including patient monitors, vital signs monitors.
  - Low risk devices, such as sequential precision devices, beds, stretchers, or tables—if fail, can be swapped out.
  - No risk devices are devices such as centrifuges, tube shakers, laboratory incubators.
- Mission critical devices (YES/NO) — Differ from site to site—user assigned.
- These are devices that, if they go down, can put patient care at risk. Some hospitals are designated as a stroke center, so the CT scanner must be up all the time. Should one of these CT scanners develop a cyber risk, it can cause a hospital to go on bypass delaying the patients' diagnoses and care as well as an affect revenue stream. If a hospital has too many bypasses for down equipment, they can lose certification and credentialing.
- Disaster recovery tiers (0–4)
  - 0–DNS server, EHR application, Billing Application, PAC's EKG storage
  - 1–3–assigned
  - 4–default to laptops, printers, etc.

Best practices used for threat detection and blocking includes:

1. Block any MRI reaching out to bad IP/URL and unwanted Geo sites to prevent phishing attacks and potential ransomware command and control centers. Absolutely no interaction with any site that has a web reputation score less than 20/30.
2. Allow predictive maintenance and performance monitoring from manufacturer web sites—specific to each model/make these accesses could be baselined and whitelisted ahead of time, only allow a specific website/URL to access.
3. Allow selective communication with other medical workstations and or server groups.

4. Stop malware spreads—Any healthcare worker laptop can inject a WannaCry type packet into an MRI running older windows without patching. Restrict internal traffic from MRI in the same segment reaching out to the corporate network.
5. Perform protocol threshold checks—excessive DNS (domain name system) requests, etc.
6. Check for embedded USB and Bluetooth drives connected to the device.

An example of threat detection and blocking is illustrated in the live software screenshots shown in Figure 6.1.

The best practices that are used by security monitoring software for tracking medical users and applications are as follows:

1. Track MRI image uploads to PACS (picture archiving and communication system) via the flow genome.
2. Check to ensure encrypted channels are used to transfer images. Mark, where there is no encryption to understand ePHI leakages.
3. Track users are accessing MRI machines with AD integration or SSO (single sign-on) platforms. Check the configuration of local users in each machine.
4. Track audit logs for user logins that are used to access the MRI machine.
5. Disable all supervisory commands like rLogin, SSH, RSH, etc.
6. Watch for any FTP/TFTP type bulk data transfer apps running on the system.

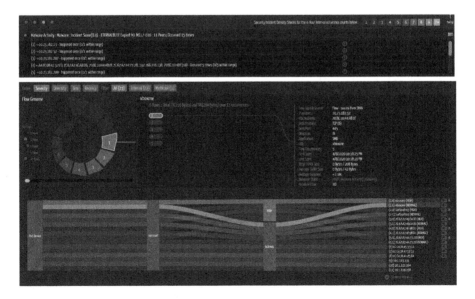

FIGURE 6.1   An example of rapid threat detection and blocking.

The best practices that security monitoring softer implements for tracking utilization for capital budget planning include:

1. Track MRI image count, study count, % of the time in use, time in the scan, etc.
2. Track the technician/users who perform scans and the type of procedure so optimization can be applied for upgrades.
3. Track the utilization of MRI usage data over a time period with thresholds (overutilized, normal, underutilized).
4. Compare and contrast multiple MRI machine for usage per location.
5. Track MRI machine cumulative image count data rather than simple time in service for setting up maintenance schedules.

6.4.3 Risk Scores

Risk Scores are computed to provide a way of categorizing what actions, if any, need to be taken with regard to individual devices.

High risks (Need to take actions, score: 7.0–8.9): Bad URL, phishing site activity (URL reputation 0–100 maps). IDS—reconnaissance attempts, malware, Trojans (CVSS score). Behavior violations (CVSS score).

Critical risks (Need immediate action, score: 9.0–10.0): External command and control communication (Critical IDS and URL reputation CVSS score). Data exfiltration (based on type and volume of data).

In order to effectively respond to various device risks, a device risk assessment framework has been developed based on the CVSS framework. Although the capabilities and features of software from different sources vary, most include the following attributes.

The framework integrates top security tools and their feedback information as well as their logs, including:

- SIEM—Splunk
- Infoblox—IPAM (IP Address Management) Info
- Palo Alto Cortex—IOCs and device data; PAN/Panorama—firewall rules
- Firewalls—Fortinet, Palo Alto networks, checkpoint
- Switches—Cisco, HPE, Extreme, H3C, Huawei, Avaya, etc.
- Wireless—CSCO, Aruba, Meraki, etc.

The framework also needs to integrate all major threat feeds, including:

- Integrated IDS leveraging signatures from major signature providers with a sophisticated threat research laboratories.
- Community IDS rules constantly developed by the open source community.
- Integrated URL/IP reputation to understand the level threats posed by the new ransomware web sites.
- IP reputation to track the reputation of external IPS.

6.5 NEW APPROACH FOR PROTECTING NETWORKED MEDICAL DEVICES

A new, innovative scheme to protect the hyperconnected hospital is required to keep track of all the IoT devices in the environment and to detect security breaches. The operation needs to be protected without inducing downtime. The system needs to be designed to identify these devices accurately in terms of their make, model, modalities, serial number, and other attributes. It must use multi-factor authentication to prevent rogue devices. It needs to map the location and topological connectivity of the building, floor, ethernet switch, wireless access point, and other network equipment.

The system must also monitor all the conversations and device transactions to detect any suspicious behavior in real-time. For example, if and when the network traffic of an MRI deviates from its regular pattern, the remedial action must be quick and automatic.

The security tools developed also identify anomalous behaviors of various devices. The baseline profiling allows immediate recognition of anomalous behaviors, preventing nefarious activities, and severely limiting potential damage and liability from a threat attack. The security tools focus on major threats, including those shown in Table 6.5.

In order to prevent gaps in security coverage, the security monitoring system must be proactive and automated. Security policies must be generated and enforced in real-time without human intervention, which means going beyond reporting alarms. In order to keep up with the number of devices, the sheer amount of traffic, and the dynamic nature of the operation, artificial intelligence (AI) and Machine Learning (ML) must be deployed.

Machine Learning (ML) allows security monitoring to detect anomalies of these devices by understanding patterns and deviations. Deep Learning allows continuous improvement inaccuracy. As healthcare becomes more dependent on networking, the security system protecting the network must become more intelligent and sophisticated.

Figure 6.2 illustrates the details of current data flows and organization of networked devices in a generic healthcare operation. As shown in Figure 6.2, each domain in the network has its own security concerns, and security threats may come from either inside the network or from the Internet.

TABLE 6.5
Threat Descriptions

Threat	Description
Phishing	Socially engineered laptop coming back into enterprise IT to spread malware.
Tampering	Replacing a badge reader with a hacker-friendly device to get into the network.
Spoofing	Weak TLS stack inpatient monitoring device to get a copy of patient data.
Denial of service	Default password to hijack a camera and launch a DDoS attack on critical assets.
Ransomware	X-ray machines with old Windows XP controlled externally for encrypting data.
Data exfiltration	Printers used as storage for data exfiltration using tunnels to C&C.

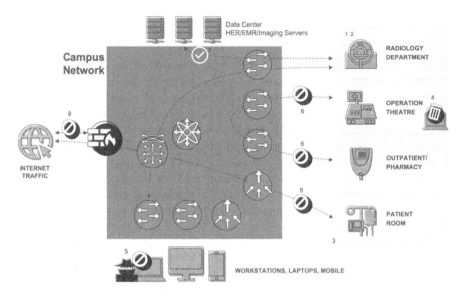

FIGURE 6.2   Architecture of medical device protection practices.

6.5.1   SOLUTIONS FOR IoT DEVICES USED IN HEALTHCARE

Figure 6.3 shows a typical implementation solution for a robust security monitoring solution. Network traffic may be monitored at strategic points with purpose-built sensors, which is then compiled into meta-data, that is, sent (usually via the same network that is being monitored) to a device analytics server. This server maintains a real-time database of device status and events, which is monitored for security information. A user interface is used to examine the data in various ways and periodic reports are generated to keep the operating staff informed.

Accurate classification and closed-loop security require mapping the complete device flow genome rather than a static device fingerprint need to do the following:

- Microscopic detail and continuous mapping of device-specific attributes.
- Rich device profile library for accurate classification.
- Real-time flow-level communications analysis.
- Advanced ML to create behavioral baselines.
- Continuous inspection to detect vulnerabilities and anomalous behavior.
- Analyzes operational performance and utilization of devices and systems.

If an event is detected by the AI software that requires immediate attention, the analytics server can take various actions. These include:

- Initiate attention of technicians by sending SMS or e-mail.
- Send instructions to firewalls to interdict suspicious traffic.

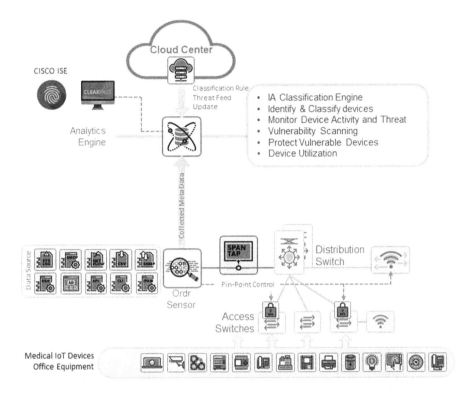

FIGURE 6.3   Architecture and cooperative cloud service.

The specific actions taken in response to specific threat profiles and levels are configurable by the system operators.

Because the nature of security threats is continuously being developed and upgraded, the analytics server subscribes to a cloud-based service. This service is cooperative in the new detection logic and parameters can be developed by examining data collected at any given site, then shared with all sites served by the same cloud service.

New systems will utilize medical device protection best practices as listed in the following sections:

DEVICE HYGIENE

1. Whitelist medical devices reconciling with CMDB (configuration management database), CMMS (computerized maintenance management system).
2. Patch vulnerable medical devices—X-rays with XP.
3. Create major partitions for medical vs. outpatient vs. contractors—segmentation.
4. Watch for wireless rogue access points, close unused open ports, update the password.
5. Block hijacked credentials to med-devices—Ransomware.

NETWORK HYGIENE

1. Control group-to-group access.
2. Whitelist flows for this imaging and EMR (electronic health record) servers to medical devices.
3. Deny anomalous flows to med-devices—micro-segmentation.
4. Block bad URL access phishing.

6.5.2 THREAT SCENARIOS

Security threats continuously evolve. Industry experts have over the years developed conceptual models that can be used to understand the nature of the threats, so that effective countermeasures can be devised. Figure 6.4 shows one threat profile in relation to an MRI system.

As shown in Figure 6.4, malware can be delivered and operate over connections that are separate from authorized connections. In this one example, it is shown how

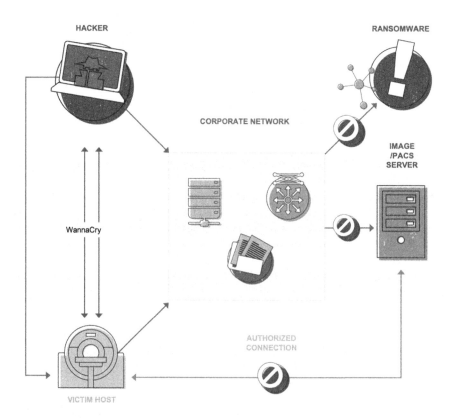

FIGURE 6.4 Security threats and events in network.

a hacker can infect an MRI (magnetic resonance imager) system via an illicit connection. The payload of the infection then may move from the MRI system to the Image/PACS server via its regularly authorized connection. In this case, the delivered malware is WannaCry ransomware, whose highest-value target is the server that has a repository of PHI (personal health information). Since the connection between the MRI system and the server is required to function, the hacker can count on that connection to move the malware laterally.

6.5.3 Third-Party Integrated Solutions for Medical IoT Devices

The solutions consist of core technologies and components that are integrated and supported by third-party tools. The integration of these third-party tools empowers the security of medical IoT devices. The basic flow and operation include:

1. Data into other systems like NAC (Network access control).
2. Policy + Manual or automated with workflow Policy generated.
3. Group membership—based on business rules in terms of department, ownership, emergency vs. regular clinic, etc., It is also important to group devices based on make, model, modalities, and other device attributes. ACL (access control list) is applied to medical where the medical devices is connected in a corporate network.

A system works with switch, firewalls, and other devices to provide all the controls required to protect it. Specifically,

1. Extracts valuable information from packets and NETFLOW;
2. provides traffic analysis based on interactions; and
3. uses access control ACLs for permit/deny.

Suspicious activities of the devices can also be detected by the system.

High level considerations for medical device protection recommended by agencies such as Department of defense in United States. These include:

- Automated network discovery and provisioning.
- Automated endpoint discovery and compliance.
- Real time visibility and tracking for reporting.
- Prevention of malware and stop data exfiltration.
- Automating remediation for breach.

In the case of medical devices, we use our Closed-Loop security for prevention and protection where access control lists, access restrictions, and Firewall rules are used. As shown in the figure, the MRI machine has conversations and communications to the corporate servers and to its peer machines only through a connection via either the wired switch or wireless access point/controller or through external firewall connections. By observing all these communications and transactions, we can assess the

TABLE 6.6

Enforcement Points for a Medical Devices

Enforcement Points	Enforcements
Wiring closet switch	Access control lists.
Wireless access points	Role based access control on their controllers.
Firewalls	Firewall rules with their TAGS.
Servers	VMware/NSX policies.
VPN access	VPN gateway access control policies.
Application tracking	Control at load balancers in the data center.
User tracking	Governance control with authentication and authorization systems.

risk involved as well as protect the device by the mechanisms listed on the right side of Table 6.6.

This is a different view of the same situation surrounding an MRI device as shown in Figure 6.4. In this case, all available enforcements are supported, affectingly isolating the MRI device from outside intrusion.

6.5.4 BEST PRACTICES FOR MEDICAL DEVICE PROTECTION

Five major best practices include:

1. DEVICE INVENTORY, UPDATES AND WHITELISTING
   - Whitelist medical devices reconciling with CMDB/CMMS.
   - Patch (OS/AV) vulnerable medical Devices; Identify older OS like XP.
   - Identify facility devices—elevator, phones to understand the interactions.
   - Update password, close open ports, vulnerability scan if possible.
2. GROUP BY DEV MODEL AND SEGMENTATION
   - Create network segments for medical vs. facilities vs. contractor vs. ER vs. pharmacy vs. guest.
   - Selectively allow group-to-group access.
3. HYGIENE ON NETWORK ENVIRONMENT
   - Plug open ethernet ports on the walls, Patient waiting areas.
   - Disconnect rogue APs.
   - Prevent devices from guest network accessing clinical resources.
   - Identify and remove move outlier devices from wrong segments.
4. SWITCH/WIRELESS POLICY FOR micro-segmentation
   - Stop malware spreads—Restrict internal traffic from devices in the same segment reaching out to medical or facility devices.
   - Whitelist internal flows for medical devices with for imaging and EMR or EHR (electronic health records) servers.
5. FIREWALL POLICY FOR EXT. COMMUNICATION
   - Block any device reaching out to bad IP/URL to prevent phishing attacks.

- Block unwanted users accessing medical devices—zero trust model with admin access.
- Ransomware—prevent medical workstations used to reach social sites.

6.5.5 ARCHITECTURE WITH MICRO-SEGMENTATION

For enterprise applications, developed and used micro-segmentation architecture as listed in Table 6.7.

Micro-segmentation can also be used for Blacklisting. Based on the conditions, security actions can be taken. Such conditions include any of:

1. Aggressive internal lateral movement using well-known IDS signatures.
2. External internet connection with bad URL reputation.
3. Massive file transfers using protocols like FTP.
4. Behavioral violation with respect to its peer group—all GE CTs vs. this GE CT (computed tomography).
5. Trending change with respect to historical patterns, last few weeks.
6. Unnecessary supervisor protocols like SSH, SCP, Telnet, Rlogin, etc.,
7. Unnecessary protocol interaction using RDP used for vendor troubleshooting.
8. Non-medical group (med and med workstations) interactions.

The actions are taken based on the conditions above, including:

1. Generate an alert immediately for SOC (security operations center) logging.
2. Automate the actions of quarantine—shut the port, blacklist MAC for wireless, assign to a quarantine VLAN, block at Firewall.

The system enables (access control list) policies. Practically speaking the process is as follows:

1. Users connect to the network, authorized passively.
2. Traffic traverses' network to the medical devices say a CT scanner.

TABLE 6.7
Understanding Attack Origin

Source Type	Description
Local L2 access	From a device connected to the same switch.
One-hop L2 access	From a device connected to a neighboring switch in the same wiring closet.
L3 access	From a device coming from the same subnet or the same floor.
Building	From a switch across the building through a distribution/core switch.
Internet	Attack coming from an external entity through a firewall.
Wireless	Attack coming from a wireless user connected to the same wireless controller.

3. A whitelisting is applied on the switch to allow or not allow that flow.
4. A monitoring system to constantly see if there are violations on the policy applied based on the base lines.

In addition, AI-based rule engines have been developed with knowledge/rule bases to classify IoT devices and detect anomalies of the devices, analyzing the data flow, and traffic behaviors.

6.5.6 Artificial Intelligence and Machine Learning

Applying ML and AI can help drive efficiency when it comes to connecting the potential billions of IoT devices to corporate networks. Automated data collection on device type and the typical traffic patterns are readily applied using various learning models. Processing the data with ML can aid network managers to cope with the numerous types of devices that are being connected, helping to deal proactively to network issues rather than the tedium of counting and classifying devices.

Automated device profiling and modeling can help understand and evaluate a network system helping to predict the appropriate behaviors of devices, enabling the system to learn and adapt as new devices are added.

Technically, AI and ML algorithms are being developed and deployed to detect anomalies of these devices by automatically understanding patterns and deviations. Human knowledge of classifying IoT devices and detecting their anomalies has been encoded into a large number of rules. IoT devices are automatically profiled by their features based on these rules that match device features such as protocols (DHCP, DICOM, etc.), manufacturers (GE, HP, etc.), and groups (medical devices, workstations, etc.).

Beyond automated classification, systems need to be grouped devices by type, class, and category, representing a hierarchical order. Details such as manufacturer, model, serial numbers, hardware, and software versions are automatically collected. Devices that share protected health information or payment card information are automatically noted as well. Sensors deployed in the network start their discovery process by inspecting the network traffic tied to the expanding profile library. AI-based classification system is being improved to analyze the traffic at multiple layers understanding the flow and not just discrete elements.

Furthermore, dealing with unknowns is remedied as the system will be connected to a large data lake to cross-reference new devices that are specific to one network which may have been encountered in another. AI and ML systems communicate and become increasingly smart to create new profiles for devices not encountered before. The larger the data lake, the smarter and faster the learning model will become.

The top ten steps to make a healthcare organization more resilient and reliable— Developing a cybersecurity framework include the following level of controls:

1. Device control—Do you know what you have:
   • Allow only whitelisted MAC addresses.
   • Allow only systems with the approved OS versions and updates.

- Allow only systems that are installed with the correct anti-virus software.
- Scan for open ports and default passwords.

2. Ownership control:
   - Identify owners, departments—This is a major data cleanup exercise but has to be undertaken to know what machines are legitimate to get connected to a healthcare network.
   - What is critical, what is mission-critical, what is clinically impacting, etc.,

3. Vulnerability control:
   - How many have FDA recall, how many manufacturing recalls, VxWORKS vulnerabilities, known issues in the firmware, older OS like Windows XP and Windows 7.
   - Onboarding process, SBOM, MDS2 forms, vendor information, to make sure there is good hygiene when a device is inducted into the network.

4. Network control:
   - Enforce VLAN/VRF/group membership and restrict movement with macro segmentation.
   - Enforce network attachment point—either a wired port or a wireless access point.
   - How mixed up are these devices with other devices in the network, how many of these devices come in close proximity with a rogue access point or a switch with open ports.

5. Traffic control:
   - IDS/IPS signature checks constantly on the traffic stream for detecting malware spreads.
   - Regulate internal lateral movement to prevent spoofing, probing, DOS attack, reconnaissance, etc.,

6. Behavior control:
   - Enforce behavior baseline with peer groups and device flow trends over last few weeks.
   - Switch port ACLs and wireless controller roles to protect these devices from other devices attempting to infect these machines.

7. Application control:
   - Application group membership and transaction enforcement.
   - Server access restriction using VMware/NSX rules.
   - Governance—which doctor is accessing which apps. How they are accessing—for e.g. through a guest access network.

8. User access control:
   - Track users to machine relationship with AD and report.
   - Identity governance with authorization systems.

9. Web access control:
   - Regulate external connections with border firewall rules to prevent command and control and ransomware.
   - Windows personal firewall.

10. Cloud services control:
   - Extend for VPN based access, branch firewalls.
   - Cloud firewall integration for mobile and remote access.

6.6   CONCLUSIONS

In this chapter, the major security challenges of IoT and medical devices in healthcare were introduced. To address these challenges, a novel architecture and software components were explained to protect a rapidly growing number of IoT and medical devices. This system is designed to identify these devices by their make, model, modality, serial number, and others. By using and analyzing the datasets from ethernet switches, wireless access points, and network sensors, the system successfully monitors all the conversations and device transactions and detect any suspicious behaviors in real-time. Rules for firewalls, access control lists for switches, role-based controls for wireless access are employed to make sure precious medical devices are protected from the internal lateral movement as well as attacks from the internet.

Security policies are generated and enforced in the system without human intervention, which means going beyond reporting alarms. More AI and ML algorithms are being incorporated in the system and deployed it to detect anomalies of these devices by automatically understanding patterns and deviations from massive device datasets and flows.

REFERENCES

1. https://www.idc.com/
2. https://www.grandviewresearch.com/

7 IoT-Based WBAN Health Monitoring System with Security

Saswati Paramita
Dayananda Sagar University
Bengaluru, Karnataka, India

CONTENTS

7.1   INTRODUCTION

Currently, the health monitoring systems have great challenges with respect to the elderly population growth rate, increase in the number of patients suffering from chronic diseases, and the rise in cost of healthcare. Millions of people are suffering from cancer, cardiovascular diseases, asthma, blood pressure, diabetes, and many other chronic diseases, these diseases get diagnosed from different symptoms at the last minute which increases the rate of the life risk. Therefore, wireless body area network (WBAN) with the application of Internet of Things (IoT) has become a ubiquitous and pervasive technology which supports continuous health monitoring system [1–2].

WBAN is one of the evolving technologies for monitoring the health conditions continuously by placing the sensor nodes on patient's body in a network. Moreover, WBAN monitors the health condition of the patient remotely independently irrespective of the physical presence of the patient at the monitoring station, i.e., anywhere at any time. However, the sensor nodes collect the information of patient's health such as ECG, blood pressure, blood glucose, and body temperature, etc. of the patient continuously, process the data and transmit to the monitoring station that is to the hospital or the doctor's end via gateways, such as mobile phone [3], PC, which are connected to the internet for monitoring the health condition of the patient [4]. Here, WBAN is integrated with the internet for pushing the collected data to the internet server [5] and communicating the patient's health condition to the monitoring station. Hence, this system behaves like a WBAN-based IoT healthcare system. After receiving the data from the internet server at the monitoring station, if there is an emergency condition detected by the doctor by analyzing the data then immediately the patient will get a call from the hospital or the smart ambulance will be connected to the patient through IoT system for the treatment [6]. But this system has an issue of security and privacy to protect patient's data from attackers [7–12]. Therefore, the patient's information should be accessed from a dedicated IoT-based WBAN network with proper authorization. Moreover, there are more challenges for security of WBANs due to mobility of the patient. As an example, there is the chance of tracking more than one sensor data of a same sensor node by an adversary even though without knowing the circumstance of the traffic [7]. Apart from this, the adversary may attack physically against the patient by obtaining the information about the critical condition of the patient.

7.1.1   WHAT IS WBAN?

WBAN is a special kind of wireless sensor network (WSN) where wearable sensors are deployed on the body of human by forming a wireless network [13]. The sensor nodes attached to the body collect different body parameters like body temperature, blood sugar level, ECG, EEG, heart rate, blood pressure, insulin level, movement, etc. The sensor nodes are basically connected in a star topology in a WBAN. However, there is a master node connected centrally with other sensor nodes known as slaves in star network. Furthermore, the central node collects the data from other sensor nodes, processes and transmits to the monitoring station via a gateway which is connected to the internet server. Therefore, a high battery power [14–16] sensor is necessary at a central node as compared to other nodes, which only collects the information by sensing the body parameters of the human being. In general, the sensor nodes which only collect the information are powered by a low power battery [16, 17].

Recent advancement of WBAN has been concentrated on a wide range of medical and non-medical applications [3–8] such as continuous remote health monitoring, immediate medical support, remote medicinal treatment, social networking, sports, information exchange, entertainment, etc. by collecting different physiological and circumstantial information using wearable or attached WBAN sensors to the human body.

7.1.2 What Is IoT?

IoT is a wireless network where several things or objects such as wearables, sensor devices, home appliances, walking machines, tablets, and smartphones, etc. are connected together. The communication is enabled between them by exchanging information using machine-to-machine (M2M) [18] communication technology without human interference.

IoT plays a great role in connecting unconnected things or objects to a network through internet for communicating with each other or people or other objects. Thus the connecting world evolves from connecting anyone, anywhere, anytime to anything. So the objects or things connected to internet are called IoT devices, which are embedded with sensors and actuators [19, 20]. Therefore, IoT technology provides the services to control or monitor anything remotely from anywhere anytime by anyone.

The IoT devices that are connected to a network through internet, collect and share the information with each other and also transfer to other network for data analysis or computation of data as per the requirement of the user. The device or network or people connected to the internet become smart for exchanging information between them or with others. However, several application areas of IoT, such as healthcare monitoring, industrial monitoring, environmental monitoring, home automation, vehicle telematics, agriculture monitoring, and smart traffic control, etc. are getting the benefit of collecting and accessing the information in real-time, which was not possible before. Furthermore, the collected information are stored in cloud and shared with the monitoring stations as per emergency alert or requirement. So, cloud computing [1] is used to control the collected data from IoT devices and provides the facilities to share resources such as integrating data service with data storage scalability, flexibility, parallel processing, and security issues. Therefore, the future prediction provides the information that about 40 billions of smart IoT devices would be connected to the internet with frequent updates to cloud by 2021 for an enormous number of IoT applications [21, 22].

7.1.3 Integration of IoT with WBAN

In the year 1995, WBAN technology accomplished a great improvement for implementing communication around or on the human body by using wireless personal

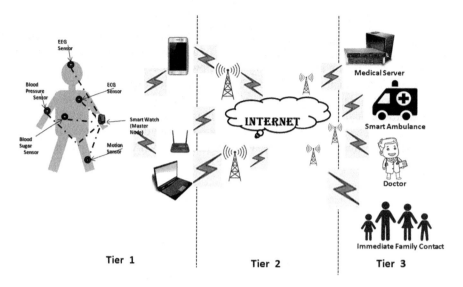

FIGURE 7.1   Three-tier architecture of IoT-Based WBAN health monitoring system.

area network (WPAN) technology. By 2001, with advancement of wireless communication technology, micro electro mechanical systems (MEMS) technology
along with integrated circuits (ICs) has empowered the use of low power [23], min-
iaturized, intelligent or smart, micro and nano technology based sensors and which
are placed around or on human body for different applications such as healthcare
monitoring, motion monitoring, etc. [23]. Furthermore, other wireless technologies
like radio frequency identification (RFID), Bluetooth, Zigbee, wireless local area
network (WLAN), internet, cellular network are also interfaced with WBAN for
communication.

The architecture as shown in Figure 7.1 follows three tiers [4] of communication, i.e., first, from master node to gateways or personal devices such as laptops
or computers, smart phones, notebooks, tablets; second, from gateways to internet;
and third, from internet to monitoring station (smart hospital [24], doctor/authorized
user, medical server, etc.). In this manner, IoT is integrated with WBAN. Here, the
personal devices act as IoT devices and the information of WBAN are accessed or
monitored through these devices with the help of internet by the monitoring station.
Intra-network communication takes place inside the WBAN whereas inter-network
communication provides smart healthcare monitoring by communicating patient's
health data to the monitoring stations remotely.

7.1.4   CHALLENGES OF USING IoT-BASED WBAN SYSTEM

The IoT-based WBAN system collects and stores the information at cloud through
the IoT devices and shares that information from the cloud [1] to the monitoring
station using internet for analysis. In this case, there is a chance of hacking the
information from the cloud or during the communication between IoT devices and

cloud or cloud and monitoring station by the attacker. Thus, the private information of a person can be accessed and used in wrong intention by the attacker. Hence, the lack of security leads to leak the personal information. Therefore, cryptography [7, 9, 12] algorithm like encryption and decryption is used to overcome such type of security issues. The sensor nodes implanted on human body in WBAN are battery-powered. Therefore, data processing happens at cloud, which leads to more energy consumption. So, IoT-based WBAN system needs more energy for cloud computing.

7.2 IoT-BASED WBAN HEALTH MONITORING SYSTEM

7.2.1 ARCHITECTURE OF IoT-BASED WBAN HEALTH MONITORING SYSTEM

The three-tier architecture [4] of IoT-based WBAN health monitoring system is illustrated in Figure 7.1.

Primarily in tier 1 communication, the WBAN sensor nodes transmit the captured data continuously to the personal server (central or master node) through Bluetooth (802.15.1 Protocol) or ZigBee (802.15.4 Protocol). The personal server includes mobile phone, laptop, personal computer (PC) or automatic robotic setup which communicates the received information to the medical server or emergency unit (smart ambulance) or family person or caretaker in audio or graphical form through the internet for monitoring or alert about the immediate treatment. In general, tier 2 communication ensures about information transfer between personal server and internet access point (AP), whereas tier 3 ensures the communication between the access point and medical server or emergency unit, etc.

Basically, three-tier architecture of IoT-based WBAN health monitoring system comprises major elements such as WBAN sensor nodes, personal server (PS), and medical server (MS).

7.2.1.1 WBAN Sensor Nodes

WBAN sensor nodes are small in size and have limited battery power [4, 23]. Therefore, they have less computing and communication capabilities as compare to master node enabled at the personal server. Each WBAN sensor node is equipped with a power supply (non-rechargeable battery), a microcontroller unit, analog-to-digital converter (ADC), signal conditioning unit, electronic sensor, and a transceiver as depicted in Figure 7.2.

Sensor: It is an embedded microchip that senses or captures the body parameter of the patient.

Power supply: It is used for providing required power to the sensor through a battery.

Signal conditioning: Amplification and filtration of sensed data take place in this unit to get the required amplitude level of data for digitization.

ADC: It converts sensed analog data to digital data for processing.

Microcontroller unit: It is used for controlling and processing the sensed data.

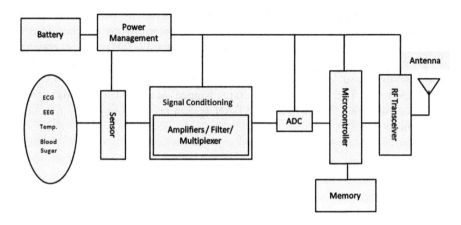

FIGURE 7.2   WBAN Sensor nodes block diagram.

Memory: It stores the sensed data temporally.
Radio transceiver: It transmits the sensed data to the PS or receives the service request or initialization command from PS.

The network configuration and management of WBAN have included comprehensively registration of sensor node to the network, initialization, customization, communication establishment.

The registration of sensor node to a network ensures the type and number of sensors used in WBAN, whereas, initialization of sensor node specifies the sampling frequency, mode of operations such as the functions during sleep time, wake up time, and emergency period, etc. Thereupon customization accomplishes the undefined calibration or signal processing. Finally, secure communication is being established by exchanging the key. This completes the configuration of WBAN network.

Afterward the network management is followed by the configuration in WBAN. The network management takes care of synchronization in time, channel sharing, data retrieval, and processing, data amalgamation. However, the PS determines the patient's health condition by analyzing the received sensed data of multiple sensors together and intimates the patient about his or her condition using a graphical or audio interface. The PS transmits the authenticated information of patient to the medical server by configuring the IP address so that the patient can be provided with necessary medical services. Furthermore, secure communication is established between the personal server and the medical server for integrating the patient's data into the medical record if there is a channel available, otherwise, the PS waits for availability of channel by storing the patient's data locally to upload the data into the medical server.

7.2.1.2 Personal Server (Master Node)
The master node in WBAN is used as a personal server [4], which acts as a gateway between the wireless sensor nodes and monitoring station (medical

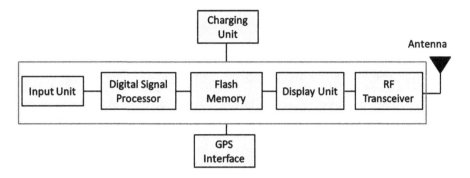

FIGURE 7.3　Block diagram of personal server.

server or emergency unit or family). The sensor nodes implanted on the body of a patient are used for capturing different health parameters. In general, smartphone or laptop or PC or automatic robotic setup is employed as PS. However, wireless sensor nodes transfer the captured parameters to the PS through a communication protocol such as ZigBee or Bluetooth. The PS analyzes the dynamic sensed information by using a signal processing unit and provides priority for transmission of the emergency information. This emergency information is transmitted to the monitoring station or emergency unit through the internet using IP address after removing redundant information or noise in the digital signal processing module of PS. The PS comprises different modules such as input unit, digital signal processor, transceiver, global positioning system (GPS) interface, flash memory, display unit, charging unit, antenna, as depicted in Figure 7.3.

7.2.1.3　Medical Server (MS)

The MS stores the received data of multiple sensor nodes in a database and frequently updated the database with continuous receiving sensed data in MS [4]. When the variation is observed in receiving data from the normal condition after data analysis, the medical unit notifies to the doctor or smart ambulance or family through the internet for immediate required action. Furthermore, MS is also accountable for the privacy of patient's health record.

7.3　IoT-BASED WBAN HEALTHCARE TECHNOLOGIES

In general, the wireless communication takes place between sensor nodes and IoT device (PS), PS to AP as well as AP to MS. Therefore, IoT-based WBAN applications depend on various design and technology requirements in wireless communication used for IoT-based WBAN healthcare system. The requirements are listed in six categories [2], such as security, energy consumption, resource management, quality of service (QoS), ubiquitous healthcare, and real-time monitoring.

7.3.1 SECURITY

In the recent advancement of information and wireless communication technologies, authentication of the system has a great role in information security as well as system security for various applications. However, a patient healthcare system must have advanced information access control techniques for sensitive and confidential patient's data along with tight security as well as maintaining the quality of data requirements. In smart healthcare system application, the physician or doctor monitors the health condition of the patient remotely by using the IoT environment in WBAN. The IoT environment facilitates the challenges in the security and privacy of the system. Therefore, cryptography algorithms [18], key agreement protocols, etc. are used for authenticating patient's sensitive data.

7.3.2 ENERGY CONSUMPTION

The IoT-based WBAN healthcare system has limited power sensor nodes used for collecting the required body parameters of the patient and transmitting to the monitoring station via the personal server. These sensor nodes consume more power for transmitting the data to the PS. Therefore, energy-efficient medium access control (MAC) protocol is used for the reduction of energy consumption by introducing sleep period or standby mode. Hence, the duty cycle is also reduced, i.e., RF antenna gets switched off when there is no relevant data for transmission. So, with a decrease in the duty cycle, the energy consumption is reduced, because,

$$\text{Duty Cycle} \propto \text{Energy Consumption} \tag{7.1}$$

$$\text{Duty Cycle} = \frac{\text{RF antenna ON time}}{\text{RF antenna ON time} + \text{RF antenna OFF time}} \tag{7.2}$$

Equation (7.1) shows the relationship between the energy consumption or the average power and duty cycle of IoT-WBAN healthcare system. The duty cycle can be reduced by controlling the sleep period with an increase of RF antenna off time as per Equation (7.2). In addition, rechargeable IoT devices are used as PS. If the patient forgets to recharge the battery of IoT device, then there is a challenge for monitoring the conditions of patient in IoT-based WBAN healthcare systems. Therefore, it is required to find an energy-awareness procedure to detect the energy constraint property of the IoT devices for the health monitoring system.

7.3.3 RESOURCE MANAGEMENT

Efficient management of the resource is required for high quality [6, 8] healthcare monitoring system. Therefore, required resource management strategies are used to access the healthcare information of the patient all over the world. As an example, smart intensive care unit (ICU) is used as temporal analysis of patient's information. Similarly, smart ambulance provides IoT base healthcare service to the patient automatically when there is a critical situation.

7.3.4 QUALITY OF SERVICE (QoS)

The quality of service (QoS) measurement [8] of IoT-based WBAN healthcare system is analyzed by meeting the required quality in terms of quality of sensed data, energy efficiency, reliability, latency, and throughput. However, the quality of the body sensor is defined by sensed data accuracy and sensitivity. Some other way, the quality of WBAN sensor is also determined from the responded sensed data quality for a query. The quality of sensed data improves by establishing security as well as privacy in IoT-based WBAN system. The improvement in the quality of data, data security, and privacy leads to the IoT-based WBAN healthcare system to achieve better lifetime and reliability.

7.3.5 UBIQUITOUS HEALTHCARE (U-HEALTHCARE)

Growth in aging population, increase in chronic diseases are mainly responsible for driving the healthcare organizations toward ubiquitous healthcare with the advancement of healthcare technology by using enormous number of body sensors, actuators to monitor the health condition of the patient remotely using IoT-based WBAN system. Even though U-healthcare [12] system provides adequate healthcare, prevention of the disease, and better care after operative to the patient; there is ampleness of challenges such as privacy and security issues. Therefore, different encryption and decryption techniques are implemented for efficient pervasive U-healthcare applications.

7.3.6 REAL TIME MONITORING

Real-time monitoring is a necessary part of remote health monitoring system [2]. Because the patient can be treated at right time as per the emergency condition and life risk may be avoided. Automated healthcare system is an example of WBAN-based IoT real-time healthcare monitoring system. This ensures that wearable sensors are automatically monitored by using IoT framework for real-time monitoring of the patient's health condition. So, it provides better management of nursing staff, cost-effective healthcare, high security, less life risk.

7.4 SECURITY IN IoT-BASED WBAN HEALTHCARE SYSTEM

With the advancement of technology, IoT-based WBAN health monitoring system provides effective and efficient emergency services to the patient by handling various medical issues, emergency notifications, and computer-assisted restored information. However, day-by-day, the numbers of user, as well as thing, are increasing in IoT-based WBAN healthcare system. Therefore, reliable communication is required for this type of system with the concern of more number of people and things. Thus, the security issues have high priority to protect the sensitive data of patient, when the data is transmitted from IoT-based WBAN to the monitoring station through the internet. The real identity of mobile IoT-based WBAN user should be protected from the public using location privacy scheme but should be accessible or traceable by the authorities as per the requirement arises for locating the physical location of IoT devices [25].

All tiers in IoT-based WBAN healthcare system must take care about the security and privacy of patient's information. Therefore, encryption is an essential technique to protect patient's sensitive information like personal health from the attacker while data transmission takes place between PS and MS. However, partial security is required within WBAN, e.g., a low power hardware encryption is used as a solution for security in ZigBee communication.

7.4.1 ISSUES AND REQUIREMENTS

In IoT-based WBAN system, the patient's health condition is diagnosed by continuous and real time monitoring. So, the protective care can be provided to the patient in emergency case by following up the health status. But there is a risk of stealing patient's information and creating harmful situation for the patient. The issues of security [9] are data modification, eavesdropping, replaying, impersonation attack.

7.4.1.1 Data Modification

The malicious attacker can access the data from source or during transmission, make the modification in data, and send to the destination. Therefore, analysis of the modified data gives the wrong result, which leads to serious consequence for patient's health monitoring. This may cause risk for patient's life.

7.4.1.2 Eavesdropping

The wireless channel is used for data transmission. Therefore, eavesdropping operation will be easy for attacker to listen to the data during transmission. Thus, patient's confidential information may be breached by the attacker which causes more dangerous attack when compared to stealing private information of the patient's health condition.

7.4.1.3 Replaying

Attacker retransmits data within the network (IoT-WBAN), which causes malfunctioning in patient's treatment. Thus, the attacker can intercept the valid original information and transmits to the actual receiver after some time for different purpose with the same goal. Therefore, other devices' behavior of the surrounding environment could be easily changed by the attackers because of interdependence relationship with their target device. Thus, the difficulty reduces for direct attacking the target devices and bypassing original mechanism of defense by malicious use of this feature.

7.4.1.4 Impersonation Attack

The sensor nodes have individual identity in WBAN and also each node has patient's identity for communication through the internet. If the attacker gets the identity of the patient then he can mislead or make spy the sensor nodes for patient's personal information.

Security, privacy, and reliability of patient's medical data should be maintained secretly, legally, and morally to mitigate the security issue in the IoT-WBAN health-care system. This can be achieved by ensuring end-to-end security of transmitted data from sensors to actuators. Thus, reliability will be improved as well as the negative effect on patient's health can be avoided. This negative effect has an impact on reliability. Since measurement of the captured data may be intentionally changed by the attacker so that the actuator will function wrongly. This leads to generate the life risk of the patient and decrease the reliability. Therefore, required access management of WBAN sensors can be ensured with advanced IoT-based technology like 6LowPAN [9] such that only authorized party can access the patient's personal information.

7.4.2 SOLUTION

The solution for the above security issues can be achieved with cryptography mechanisms by maintaining confidentiality, privacy, and reliability. There are two types of cryptography mechanism such as symmetric and asymmetric. In symmetric cryptography mechanism, a public key is shared between the sender and receiver, whereas, in asymmetric cryptography, each end has two keys, i.e., public key and private key for communication. The public key defines the attributes of communication between two parties. But private key maintained secrecy and is used to decrypt the information from encrypted data (encrypted by public key).

7.5 SECURITY METRICS

All communication in IoT-based WBAN or exchange of data between the sender and receiver is encrypted by sharing security key [9]. Hence, unknown sensor nodes will not get the permission to access the server unless the link establishment happens by sharing the security key.

Any outside node from the IP host cannot communicate with a node in WBAN directly without the permission of the server. Therefore, all the nodes should pass the information through the server. Thus, authorized one can only communicate with sensor nodes of the WBAN network, and each node should contain their own sensed information.

The nodes identity (Id) is generated by itself. So, there is not required to share the network node Id's. Hence, the WBAN node's information cannot be accessed by sniffing attacks.

7.6 SUMMARY

IoT-based WBAN system changed the way people think about the healthcare management system just like people communicate with each other using internet and search for information independent of place and distance. This motivates for proactive and preventive healthcare with the improvement of quality of life and low-cost healthcare.

Integrated IoT-WBAN system with 3-tier architecture was discussed in this chapter for remote healthcare monitoring of patient with security challenge. However, this integrated system facilitates inter-human (human-human), inter-device (device-device), and device-to-device communication. Therefore, this is expected wireless sensor network integration for healthcare applications by experiencing the paradigm improvement in communication technology.

IoT-based WBAN health monitoring system is facing security challenges with increase in intelligent IoT devices and various applications. In addition, the patient's information should be maintained secret from attackers because the attacker may use this data illegally against the patient. Therefore, privacy is the major concern for this system.

The existing IoT-based WBAN system provides scalable, reliable, and effective health monitoring for patient by using intelligent and smart sensor with advanced communication technologies. Furthermore, the improvement of security, flexibility, and power consumption is required in this advanced system.

Future work will be focused on employing a private and secured cloud network for IoT-based WBAN healthcare system. Because employing data encryption method for privacy is a typical task for low storage sensor nodes or IoT devices. Therefore, cloud storage is used for large data storage and improving privacy of patient's personal data. Hence, the necessary IoT-based WBAN health monitoring system design is a major concern to achieve considerable efficiency, reliability, low power consumption, and high privacy.

REFERENCES

1. Sureshkumar Selvaraj, Suresh Sundara Varadhan, 2020, Challenges and opportunities in IoT healthcare systems: A systematic review, *SN Applied Sciences*, Springer Nature Journal.
2. Mrinai M. Dhanvijay, Shailaja C. Patil, 2019, Internet of Things: A survey of enabling technologies in healthcare and its applications, *Computer Networks*, Elsevier.
3. Suree Funilkul, Nipon Charoenkitkarn, Prasert Kanthamanon Debajyoti Pal, 2018, Internet-of-Things and smart homes for elderly healthcare: An end user perspective, *IEEE Access*.
4. G. Elhayatmy, Nilanjan Dey, Amira S. Ashou, 2018, *Internet of Things based wireless body area network in healthcare*, Springer International Publishing AG.
5. Subhajit Chatterjee, Shreya Chatterjee, Soumyadeep Choudhury, Sayan Basak, Srijan Dey, Suparna Sain, Kali Shreyo Ghosal, Niket Dalmia, Sachet Sircar, 2017, *Internet of Things and body area network-an integrated future*, IEEE.
6. Taiyang Wu, Fan Wu, Jean-Michel Redouté, Mehmet Rasit Yuce, 2017, An autonomous wireless body area network implementation towards IoT connected healthcare applications, *IEEE Access*, vol. 5, pp. 11413–11422.
7. Pandi Vijayakumar, Mohammad S. Obaidat, Maria Azees, S K Hafizul Islam, Neeraj Kumar, 2020, Efficient and secure anonymous authentication with location privacy for IoT-based WBANs", *IEEE Trans. Industr. Inform*, vol. 16, pp. 2603–2611.
8. Guofa Cai, Yi Fang, Jinming Wen, Guojun Han, and Xiaodong Yang, 2019, QoS-aware buffer-aided relaying implant WBAN for healthcare IoT: Opportunities and challenges, *IEEE Network*, vol. 33, pp. 96–103.
9. Anass Rghioui, Aziza L'Aarje, Fatiha Elouaai, Mohammed Bouhorma, 2014, The Internet of Things for healthcare monitoring: Security review and proposed solution, *IEEE*, pp. 384–389. https://ieeexplore.ieee.org/document/7016651.

10. Mutaz Elradi S. Saeed, Qun-Ying Liu, Guiyun Tian, Bin Gao, Fagen Li, 2018, Remote authentication schemes for wireless body area networks based on the Internet of Things, *IEEE Internet Things J.*, vol. 5, pp. 4926–4944.
11. E. Luo, M. Z. A. Bhuiyan, G. Wang, M. A. Rahman, J. Wu, M. Atiquzzaman, Feb. 2018, Privacy protector: Privacy-protected patient data collection in IoT-based health-care systems," *IEEE Commun. Mag.*, vol. 56, no. 2, pp. 163–168.
12. Ebrahim AL Alkeem, Chan Yeob Yeun, M. Jamal Zemerly, 2015, *Security and privacy framework for ubiquitous healthcare IoT devices*, IEEE, pp.70–75.
13. Aishwariya.Aetal., 2018. A survey of study on WBAN architecture for biomedical and scientific applications. *IOSR J Eng.*, vol. 08, no. 12, pp. 72–80.
14. S. Kim, S. Kim, D. S. Eom, 2013, RSSI/LQI-based transmission power control for body area networks in healthcare environment, *IEEE J Biomed. Health Inform.*, vol. 17, no. 3, pp. 561–571.
15. Stefan Mijovic, Riccardo Cavallari, Chiara Buratti, 2015, *Experimental characterization of energy consumption in body area networks*, IEEE, pp. 514–519.
16. Jitumani Sarma, Akash Katiyar, Rakesh Biswas, Hemanta Kumar Mondal, 2019, Power-aware IoT based smart health monitoring using wireless body area network, *IEEE 20th International Symposium on Quality Electronic Design*, IEEE, pp. 117–122.
17. Jamil. Y. Khan, Mehmet R. Yuce, 2010, Wireless body area network (WBAN) for medical applications, *New developments in biomedical engineering*, Domenico Campolo (Ed.), ISBN: 978-953-7619-57-2, InTech.
18. Changyan Yi, Jun Cai, 2019, A truthful mechanism for scheduling delay-constrained wireless transmissions in IoT-based healthcare networks, *IEEE Transactions On Wireless Communications*, vol. 18, no. 2, pp. 912–925.
19. T. Jagannadha Swamy, T. N. Murthy, 2019, eSmart: An IoT based Intelligent Health Monitoring and Management System for Mankind, IEEE International Conference on Computer Communication and Informatics.
20. S. Ananth, P. Sathya, Mohan P. Madhan, 2019, Smart health monitoring system through IoT. IEEE International Conference on Communication and Signal Processing, pp. 0968–0970.
21. Y. Yang, X. Zheng, W. Guo, X. Liu, V. Chang, 2018, Privacy-preserving fusion of IoT and big data for e-health. *Future Generation Computing System*, vol. 86, pp. 1437–1455.
22. P. Karthikeyyan, S. Velliangiri and M. I. T. Joseph. S, Review of Blockchain based IoT application and its security issues, 2019 2nd International Conference on Intelligent Computing, Instrumentation and Control Technologies (ICICICT), Kannur, Kerala, India, 2019, pp. 6–11, doi: 10.1109/ICICICT46008.2019.8993124.
23. A. Srilakshmi, P. Mohanapriya, D. Harini, K. Geetha, 2019, IoT based Smart Health Care System to Prevent Security Attacks in SDN, IEEE Fifth International Conference on Electrical Energy Systems, 2019, pp. 1–7.
24. https://patientengagementhit.com/features/smart-hospitals-making-the-future-of-patient-experience-a-reality.
25. S. Velliangiri, P. Karthikeyan, I. T. Joseph and S. A. P. Kumar, "Investigation of Deep Learning Schemes in Medical Application," 2019 International Conference on Computational Intelligence and Knowledge Economy (ICCIKE), Dubai, United Arab Emirates, 2019, pp. 87–92, doi: 10.1109/ICCIKE47802.2019.9004238.

8 Integration of Blockchain into IoT

J. Premalatha and K. Sathya
Kongu Engineering College
Erode, Perundurai, Tamil Nadu, India

CONTENTS

8.1   INTRODUCTION: BACKGROUND ON BLOCKCHAIN AND IoT

Blockchain is a distributed network with all nodes holding a consistent copy of immutable records of transactions. With immutable records created through some mathematical functions, it becomes highly impossible to forge any transactions in the network. Internet of Things (IoT) connects various physical devices to collect data, process it, and take actions in a controlled environment. This chapter focuses on various security limitations of IoT and usage of blockchain to overcome those issues and make them favourable for real-time applications.

8.2   IoT SECURITY LIMITATIONS

While implementing IoT, it is necessary that security be the main concern as the actions are taken intelligently by the system rather by human. In this section, we will consider the key security requirements, limitations in achieving the requirements [1].

8.2.1   SECURITY REQUIREMENTS

8.2.1.1   Data Confidentiality

Data confidentiality refers to the privacy of the data. In IoT applications concerning monetary data, other business-sensitive data, it is necessary that these data are kept secret from the outside world while they are handled in IoT. Such applications require that IoT provides confidentiality for all data that are being collected and transmitted, possibly avoiding attackers from getting hold of sensitive data.

8.2.1.2   Data Integrity

Data integrity refers to the preservation of data without modification. IoT applications in decision support systems, industrial control can be devastating if their actions are based on false data. Integrity of data requires that IoT safeguards the data from any injection attacks and any modifications.

8.2.1.3   Data Availability

Data availability refers to the availability of data whenever is required. Crucial applications like automated vehicular monitoring, manufacturing industries rely on the real-time data. When such real-time data becomes unavailable due to any system failure or attack, a catastrophic event may occur. This requires that IoT ensures safe availability of data all the time despite any attempt to make the system fail.

8.2.1.4   Authenticity

Device and user authentication are necessary to verify the identity of entities (users, devices, etc.), accessing the IoT network. Authentication refrains suspicious entities from entering the network.

8.2.1.5   Authorization

Any entity accessing the data needs to be verified for their authorization. Only entities entitled to access data are allowed to gain access, and any unauthorized entity is abstained from accessing data in IoT.

8.2.2   SECURITY CHALLENGES IN IoT

Though IoT enables communication between devices, automating the systems, being economical and efficient, users are still concerned about its security. There are few cases like hacking the vulnerabilities of smart TVs, IoT-enabled cash machines. These incidents made the IoT devices challenging to trust. This section lists the security challenges that need to be considered in future implementation [2, 3].

- *Outdated devices*—Most of the IoT devices don't get enough updates from their manufacturer, thus they are prone to attack with new techniques.
- *Default credentials*—Many vendors ship their devices with default credentials, the users install the devices unchanged. Default known credentials open gates for hackers.
- *Ransomware*—There may be situation where a hacker may gain access to the vulnerable devices and collect data in demand for ransom to return the data.
- *Secure communication*—To provide data integrity and confidentiality, the network carrying those data must be secured with the deployment of necessary cryptographic algorithms.
- *Guaranteed availability*—Data required for decision making, analysing must be available all the time despite the system failure.
- *Single point of failure*—All the devices are connected with an external server where data are analysed and actions are taken. Any failure of the server results in devices being not able to communicate.

8.3 BLOCKCHAIN

Blockchain is an open, decentralized network where every transaction is validated by every node in the network [4]. By this way, no node can forge a transaction as all the nodes in the network validate all the transactions, and stores all the transactions. When the number of transactions reaches a limit, they are bundled into a block, and miners compute a hash value with certain degree of difficulty for a block. The new block with computed hash value is dispersed to every node in the network and added to the existing chain of blocks (Blockchain). The link is made when the hash value of the new block is computed from the set of transactions and hash value of the previous block. Linking the blocks this way guarantees that the transactions cannot be forged as the hash value of all the subsequent blocks gets modified.

The miners play a key role in computing the hash value for each block. The hash value computed has a higher degree of complexity with a predefined number of zeros in prefix/suffix. Predefining the difficulty level guarantees that whenever a change is made to the blockchain, it is impossible to create a new hash value satisfying the predefined condition. This safeguard against the forging of transactions is added into the blockchain. Some examples of transactions may include money transfer between accounts, transfer of data from one person to another, transfer of business data between parties, etc. Figure 8.1 depicts the entire lifecycle of a transaction in a blockchain.

8.3.1 DISTRIBUTED LEDGER

A digital ledger was developed to make note of all transactions and their credit/debit amount in monetary terms. Blockchain uses a similar structure to record all the transactions in the network with details of identity of creditor, identity of debtor, time of transaction, amount of money transferred. These details are stored in the

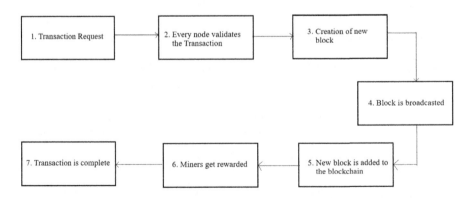

FIGURE 8.1   Lifecycle of a transaction in a blockchain.

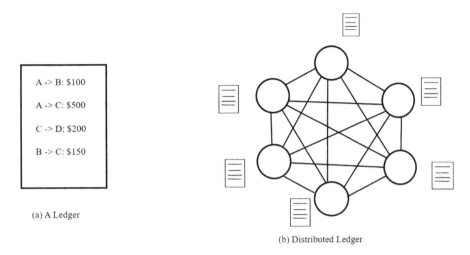

(a) A Ledger

(b) Distributed Ledger

FIGURE 8.2   (a) A ledger. (b) Distributed ledger.

blocks. Hence the blocks in blockchain are called ledgers. The blocks once validated are broadcasted to all nodes in the network so as all nodes have a local copy of all validated blocks. This coins the term "Distributed Ledger" as a single ledger is consistently distributed among all nodes of the network. All nodes with a consistent copy of distributed ledger are shown in Figure 8.2.

8.3.2   BLOCKCHAIN CONSENSUS

The working of blockchain goes well until only when all the peers have the same copy of ledger with them. In practical world with peers distributed across the globe, it is always highly difficult to have the same copy of the ledger in all nodes. Consensus is a process where a group of people makes a common agreement on any decision without a leader or convenor. Since blockchain is decentralized with no central authority to control, the participating nodes have to make a consensus on making all decisions. Consensus protocols make sure that the new block added to the blockchain is the only true copy, that is, being added in all peers distributed ledger. Consensus protocol mandates equal participation of all nodes in the process of coming to an agreement. This way, all nodes are given an equal chance to be a miner. Some of the consensus protocols are:

Proof of work (PoW)
Proof of stake (PoS)
Proof of burn (PoB)
Proof of capacity
Proof of elapsed time

8.3.3 Security Features of Blockchain

- *Decentralization of authority*—No centralized single node holds the authority to validate all the transactions [5, 6].
- *Peer-to-peer network*—P2P network allows two nodes in a network to communicate directly without a need for any trusted third-party.
- *Resistance to tampering*—The use of hash value enables the verification of the integrity of a block. A hacker trying to change a transaction in a block has to change all the hash values of all subsequent blocks satisfying the predefined difficulty level. It is computationally infeasible to perform such change, thus safeguarding the chain of blocks against mutations.

8.4 APPLICATIONS OF IoT DEMANDING HIGH SECURITY

As we have discussed, IoT is aiming to make the physical devices smarter enough to act intelligently without any human interference. Hewlett Packard (HP) surveyed the population of devices connected through IoT, and they made an estimation of devices population in the coming years as shown in Table 8.1 [7].

Many applications store their data in a secure cloud server, where the data analytics takes place, and decisions are made. Application builders are now burdened with securing the data while in transit while it is stored in the cloud.

Some of the IoT applications that have become existent with our everyday life requiring a certain level of security are listed below [8]:

- Smart home
- Smart wearables
- Smart cities
- Industrial IoT
- Internet of Military Things (IoMT)
- Ocean of Things

TABLE 8.1

Survey on IoT Devices

Year	Physical Devices Connected to IoT
1990	300,000
1999	90 million
2010	5 billion
2013	9 billion
2025	1 trillion

Note: (Data from HP, 2014)

8.5   BLOCKCHAIN—THE SOLUTION FOR IOT'S DEMAND

In this section, the security features demanded by IoT are listed, and how the demands can be satisfied by the use of blockchain is also discussed [9, 10].

8.5.1   IoT Demands

Processing of data cannot be done on the smart physical devices due to limited resources of memory, power, and processing capacity. This forces the IoT to push the gathered data onto the cloud for processing and analysing to take actions. In this scenario, IoT has three demands which are:

- *Data collection from heterogeneous systems*—The IoT in any application is ought to have interconnected heterogeneous systems, each with different functions. This requires the devices to collect real-time data from their environment and deliver the data to the cloud seamlessly with less effort and less cost.
- *Security of collected data*—IoT requires the collected data to be resistant against any tampers. Any change in data can lead to a change in decision taken, which may incur heavy loss or threat to the application.
- *Ability to share data across multiple parties*—Today, IoT applications built for large organizations typically involves multiple parties. All the parties have the rights to access the data and validate the decisions made.

8.5.2   Blockchain—The Solution

Blockchain has the following features that help to provide a solution to the demands of IoT stated above [11]:

- *Create transactions for any activity on the network*—Blockchain can be designed to record any non-monetary actions taking place in a system, for example, real-time movement of shipments and can be secured as proof for future reference in case of lost or misguided shipments.
- *Blocks are tamper-proof and accessed only by authorized peer nodes*— Blocks once get verified by peer nodes are mined by miners. Mined blocks get their unique hash value and get distributed in the network. Hash technique makes the blocks resistant to any tampering attacks.
- *Blocks are distributed to all peers in the network consistently*— Consensus algorithms aim to maintain a consistent copy of the distributed ledger (chain of blocks) at all authorized peer nodes. When a node becomes unavailable due to reasons, like out of coverage of the network, system failure, it gets a recent consistent copy of blockchain from its neighbours after it rejoins the network.

As we have discussed, blockchain seems to be an ideal solution for IoT applications where a secure collection of data is required. In the next section, we will review how to integrate blockchain into an IoT application and what steps need to be carried out before integration.

8.6   INTEGRATING BLOCKCHAIN INTO IoT—AN OPEN CHALLENGE

Blockchain as a decentralized network of nodes has its own advantageous features with more tamper-proof security and providing anonymity to users [12, 13]. However, bringing this giant concept of blockchain into IoT will certainly create certain challenges as listed below:

8.6.1   SIZE OF BLOCKCHAIN AND SCALABILITY

We know that in a blockchain, every action or transition of information is considered to be a transaction. In IoT, devices tend to collect data and transmit them to the central cloud. Practically the transmissions of data referred to as transactions are built as blocks and keep on added to the chain of blocks. Typically, a block creation incurs overhead in mining, storage space for blockchain, and broadcasting of the chain in the network. When it comes to IoT, devices have limited power and storage space that creates a challenge to deploy blockchain.

8.6.2   USER PRIVACY AND IDENTITY

Blockchain handles all the transactions openly in the network, meaning the transaction is visible to every node in the network. The transactions in IoT may contain sensitive data that need both confidentiality and integrity. To enable privacy of the transaction, data in IoT solutions need to be found that protects the identity and privacy of the IoT devices.

8.6.3   VULNERABILITY OF SMART CONTRACTS

Smart contracts are self-executing codes that are stored in blockchain. They are executed so as to enforce the rules of a common agreement between transacting parties. A smart contract is provided with a unique address and is stored in the network. When transacting parties want to enforce the agreement, send the transaction to the smart contract address, and the transaction is validated based on agreement. Since smart contracts are self-executing, they possess the vulnerability of spreading virus, malicious code to nullify or modify a transaction, bugs while creation, hacking. It is a challenge to ensure the correct functioning of a smart contract.

8.6.4   CONSENSUS PROBLEM

PoW consensus led to miners forming mining farms with a large pool of computing resources and always winning the consensus and getting rewards. This makes the system to be more centralized rather than being decentralized. Proof of stake (PoS) algorithm wants the miners to deposit a certain amount as stack for security, and the one with large stacked money wins the mining. This way, miners are prevented from acting maliciously as their money is at stack if validation goes incorrect [14–16].

8.6.5 REDUCED THROUGHPUT

Throughput of block generation in IoT devices is reduced as IoT devices are not a close match to blockchain miner devices. The devices may also frequently face communication breakdown, and this imposes a further challenge in maintaining the consistency of the blockchain.

8.7 DESIGN ISSUES IN INTEGRATION

While integrating blockchain into IoT applications, we may face practical issues in designing blockchain-based IoT applications. This section provides an insight into the design issues that need to be resolved while implementing blockchain-based IoT.

8.7.1 IDENTIFYING THE BLOCKCHAIN ENDPOINTS

Endpoint in the blockchain refers to the peer or node in the network. It is necessary to choose these endpoints in IoT that are capable of storing the blocks and validating the transactions. It is critical to choose endpoints in IoT that have enough bandwidth utilization, computing power, and storage space.

8.7.2 DETERMINING WHAT DATA TO BE ON THE BLOCKCHAIN

Any business application has its own business logic and business data. Integrating blockchain into IoT business applications, it is necessary to implement the business logic in smart contracts that get executed automatically to enforce the business rules. It is important to consider which part of business logic and business data are to be kept inside the blockchain and which to be kept out of blockchain.

8.7.3 USER IDENTITY PRIVACY ISSUE

Leakage of user identity will be a serious issue as it enables anyone to track all transactions performed by that user. User privacy is of great concern when integrating blockchain into IoT. Various known techniques like analysing user behaviour, knowing your customer, phishing can lead to invasion of user privacy.

8.7.4 PRIVACY OF TRANSACTION

The transaction in an IoT application can contain sensitive data that requires confidentiality. Many advanced techniques are available to guess and trace the transaction data by various analysis tools. Steps must be taken to protect the data inside a transaction as the block gets broadcasted in the network.

8.8   SOLUTION FOR THE ISSUES

In the previous section, the issues in designing a blockchain integrated IoT were reviewed. It is clear from the fact that these issues don't exist while discussing the benefits of integration in theory. Issues arise only during the practical implementation. It is wiser to resolve these issues in the design phase beforehand of implementation.

8.8.1   BLOCKCHAIN ENDPOINTS

The endpoint or node to be deployed in a blockchain must have enough processing power, storage space to perform mining and storing the distributed ledger. IoT devices are smaller in size, that is, capable of mostly performing a single task. To overcome this issue, the blockchain endpoint is generally any computer or any cloud-rented server, or enterprise server. IoT devices are the clients that create transactions and send them to the endpoint where validation, mining, and addition to blockchain occur. Another approach is to build Pseudo Distributed Things architecture, where each IoT device is attached with an endpoint called smart thing and are controlled by the endpoint container component present in the cloud.

8.8.2   STORING BUSINESS LOGIC AND DATA

Any application in any field is likely to have its own business logic and sensitive business data. Third-party application developers can extend the available smart contract to leverage the device features and create a new smart contract with his/her business logic. To protect the data from unauthorized access, access controls can be placed on smart contracts. Smart contracts use digitized tokens of users to verify their identity and access rights. When a new user enters the network, authentication is performed and provided with a digital token for future accesses.

8.8.3   USER PRIVACY

To avoid traffic analysis, it is encouraged to use a new key for every new transaction or at least a new key for every new receiver of the transaction. This makes it difficult to trace back the public key to a user.

8.8.4   TRANSACTIONAL PRIVACY

To guarantee the privacy of transactions, homomorphic encryption, and zero-knowledge proof can be utilized. Homomorphic encryption enables computation like mining to be performed on the cipher text without the need for plain text. Zero-knowledge proof is a technique that verifies the transaction statement without even exposing the content of the transaction.

8.9   FEASIBILITY ANALYSIS OF INTEGRATION INTO YOUR APPLICATION

Before starting to implement the integration process for your IoT application, it is necessary to perform the feasibility analysis to check whether the proposed integration of blockchain into your application is going to reap the benefits of blockchain.

8.9.1   SCALABILITY ISSUES

In order to integrate blockchain into your system, each IoT device must be treated as a unique user of blockchain network. Each data transmission is considered as a transaction. Consider a simple scenario of home security system with IoT devices. All users of the home connect all the devices to blockchain. Whenever there is a data transfer from sensor, it goes through the blockchain and notifies the users in real-time. In a practical situation, the system has to deal with hundreds of thousands of users and data transfers at the rate of tens of thousands of transactions per second. The blockchain to be integrated must be scalable to handle such voluminous transactions. The more popular blockchain network Ethereum is capable of processing only six transactions per second. It is still economically not feasible to jump from six to tens of thousands of transactions per second without an improved scalable solution.

8.9.2   SMART CONTRACT ISSUE

Smart contract in blockchain is an executable code designed to reflect the business logic. During the design phase, it is necessary to consider what part of your business logic needs to be controlled in blockchain environment. Determine this is a crucial part as the smart contract executes automatically upon specified conditions. Feasibility analysis for your application must consider what needs to go into smart contract, how they are accessed, and who can access it.

8.10   SECURITY ANALYSIS OF INTEGRATED SYSTEM

This section gives insight into the security advantages of integrating blockchain into IoT application. As blockchain itself, a decentralized network with immutable records provides various security features. Following are the security features that one can expect from integration. Also, depending on the requirement of specific features, the integration of blockchain can be designed.

8.10.1   RESILIENCE

Resilience of a system is its ability to recover from any attack or damage. Various IoT applications have their resilience with the implementation of intrusion detection. Integrating blockchain into IoT systems improves the efficient working intrusion detection algorithms by running parallel on all nodes with blockchain data as input.

8.10.2 ACCESS CONTROL

Access control is the mechanism of controlling the access to crucial resources based on the authorization of accessing node. Access control is essential when there is a sharing of resources among multiple nodes. Blockchain with IoT provides such controlled access to crucial resources, like smart contracts, smart devices, services, and IoT data. Nodes that require access to a resource need to prove that it is authorized to access the resource.

8.10.3 PRIVACY

IoT devices that integrate blockchain also require privacy for their users (devices) and their data. Devices are addressed using their public keys in all the transactions, thus the devices are anonymous to each other. This technique called pseudo-anonymous is not fully secure as it can be easily traced back to identify the device. Full anonymity can be obtained with advanced cryptographic techniques. Linkable ring signatures can be used by a sender to sign a transaction and hide the sender's identity in a spontaneous ring. Transaction privacy is achieved by homomorphic commitments or zero-knowledge protocols.

8.10.4 SECURITY USE CASES

Blockchain can be deployed in IoT for any of one the following scenario:

1. Device authentication in mesh networks.
2. Secure IoT transactions in mesh networks.
3. Chain of trust in IoT.
4. Immutable transaction records in IoT.

8.11 PERFORMANCE OF INTEGRATED SYSTEMS

Once the blockchain has been integrated into the IoT application, it is necessary to ensure that the performance of the system does not degrade. Performance of the blockchain-enabled IoT can be evaluated on the basis of two metrics, namely Average Latency and Throughput.

8.11.1 AVERAGE LATENCY

Latency refers to the waiting time of a transaction between its time of selection for validation and time of addition to a block. Blockchain performance is measured based on the average latency of all transactions. The number of transactions processed has a direct impact on the average latency. With IoT being a generator of tens of thousands of transactions, care must be taken to improve the average latency. Increased latency is not endured in real-time applications, where transactions need to be processed as soon as they are generated.

8.11.2 THROUGHPUT

Throughput is defined as the number of transactions processed and completed in unit time. Throughput of the typical Ethereum blockchain network is slightly over 14 TPS (Transactions per Second). IoT applications deploying blockchain need to consider this measure to efficiently utilize the resources in processing the transaction. Some proposed to use side chains that can process the transactions in parallel and increase the throughput (e.g., Plasma).

8.11.3 ENERGY CONSUMPTION

IoT devices implementing blockchain are power-constrained devices that generally run on batteries or limited power supply. Validating the transactions requires a considerable amount of energy. Designers must take into consideration the energy limit of those physical devices and can deploy rented computing resources for transaction validation.

8.12 TRADEOFFS BETWEEN PERFORMANCE AND SECURITY

Before starting to implement, it is wiser to make comparison and perform tradeoff between performance and security of your system. Tradeoff can be made based on the requirement of the application. IoT, with its increased number of transactions per second places a huge overload on the devices for validating them. If performance is the main concern, security must be compromised a bit since executing the security protocols also includes the overhead in computation. Real-time applications expect no delay between submission and completion of a transaction. In such cases, security implementation may introduce delays that are unacceptable.

Highly secure IoT applications provide both user privacy and data privacy. Blockchain has its own consensus protocols and mining procedures. Cryptographic encryption algorithms are deployed for the confidentiality of the transaction. Such mining, consensus, encryption processes incur more time and resources. This will inherently reduce the performance of the system.

Designer of your application must make a choice to have a balance between performance and security. This enables efficient integration of blockchain into available resources and provides the required level of security.

REFERENCES

1. Oh, Se-Ra, and Young-Gab Kim. 2017. "Security requirements analysis for the IoT." 2017 International Conference on Platform Technology and Service (PlatCon), IEEE, Busan, South Korea, pp. 1–6.
2. Panarello, Alfonso, Nachiket Tapas, Giovanni Merlino, Francesco Longo, and Antonio Puliafito. 2018. Blockchain and IoT integration: A systematic survey. *Journal of Sensors* 18 (8):2575.
3. Hang, Lei, and Do-Hyeun Kim. 2019. Design and implementation of an integrated IoT blockchain platform for sensing data integrity. *Journal of Sensors* 19 (10):2228.
4. Drescher, Daniel. 2017. *Blockchain basics.* Vol. 276. Springer.

5. Stephen, Remya, and Aneena Alex. 2018. "A Review on BlockChain Security." IOP Conference Series: Materials Science and Engineering, Kerala, India, 2018, IOP publishing.

6. Wüst, Karl, and Arthur Gervais. 2018. "Do you need a blockchain?" 2018 Crypto Valley Conference on Blockchain Technology (CVCBT), IEEE, Zug, Switzerland, pp. 45–54.

7. https://www8.hp.com/us/en/hp-news/press-release.html?id=1744676.

8. 2020. " https://mitechnews.com/internet-of-things/how-big-is-iot-20-6-billion-connected-devices-by-2020/.

9. Lo, Sin Kuang, Yue Liu, Su Yen Chia, Xiwei Xu, Qinghua Lu, Liming Zhu, and Huansheng Ning. 2019. Analysis of blockchain solutions for IoT: A systematic literature review. *IEEE Access* 7:58822–58835.

10. Maroufi, Mohammad, Reza Abdolee, and Behzad Mozaffari Tazekand. 2019. On the convergence of blockchain and Internet of Things (IoT) technologies. *Journal of Strategic Innovation and Sustainability for Issue* 14:1–11.

11. Zhang, Rui, Rui Xue, and Ling Liu. 2019. Security and privacy on blockchain. *ACM Computing Surveys* 52 (3):1–34.

12. Dai, Hong-Ning, Zibin Zheng, and Yan Zhang. 2019. Blockchain for internet of things: A survey. *IEEE Internet of Things Journal* 6 (5):8076–8094.

13. Banafa, Ahmed. 2018. "Nine IoT Predictions for 2019." https://iot.ieee.org/conferences-events/wf-iot-2014-videos/52-newsletter/november-2018.html.

14. Karthikeyyan, P., S. Velliangiri and M. I. T. Joseph S. 2019. "Review of Blockchain based IoT application and its security issues," 2019 2nd International Conference on Intelligent Computing, Instrumentation and Control Technologies (ICICICT), Kannur, Kerala, India, pp. 6–11, doi: 10.1109/ICICICT46008.2019.8993124.

15. Velliangiri, S., and D. P. Karthikeyan. 2020. "Blockchain Technology: Challenges and Security issues in Consensus algorithm," 2020 International Conference on Computer Communication and Informatics (ICCCI), Coimbatore, India, pp. 1–8, doi: 10.1109/ICCCI48352.2020.9104132.

16. Jeyabharathi, J., S. Velliangiri, Ahamed, N.N. and Karthikeyan, P., 2020. A heuristic search method unified in blockchain expertise for supply chain management. *International Journal of Control and Automation* 13(4):668–675.

9 Industrial Internet of Things Safety and Security

J. Premalatha and Vani Rajasekar
Kongu Engineering College
Erode, Perundurai, Tamil Nadu, India

CONTENTS

9.1 INTRODUCTION TO INDUSTRIAL INTERNET OF THINGS

Internet of Things (IoT) implemented in the industrial and manufacturing space, it becomes Industrial Internet of Things (IIoT). This technique is an amalgamation of various technologies like big data, machine learning, machine-to-machine communication, sensor data, and automation that have involved in the industrial backdrop for past years and general architecture is shown in Figure 9.1.

9.1.1 WHAT IS INDUSTRIAL INTERNET OF THINGS?

IIoT makes a connected enterprise by combining the operational and information department of the industry. Thereby improves visibility, efficiency, increases productivity, and thereby decreasing the complexity of the process in the industry. IIoT sometimes referred as transformative manufacturing strategy shows improved quality, productivity, and safety in an industry.

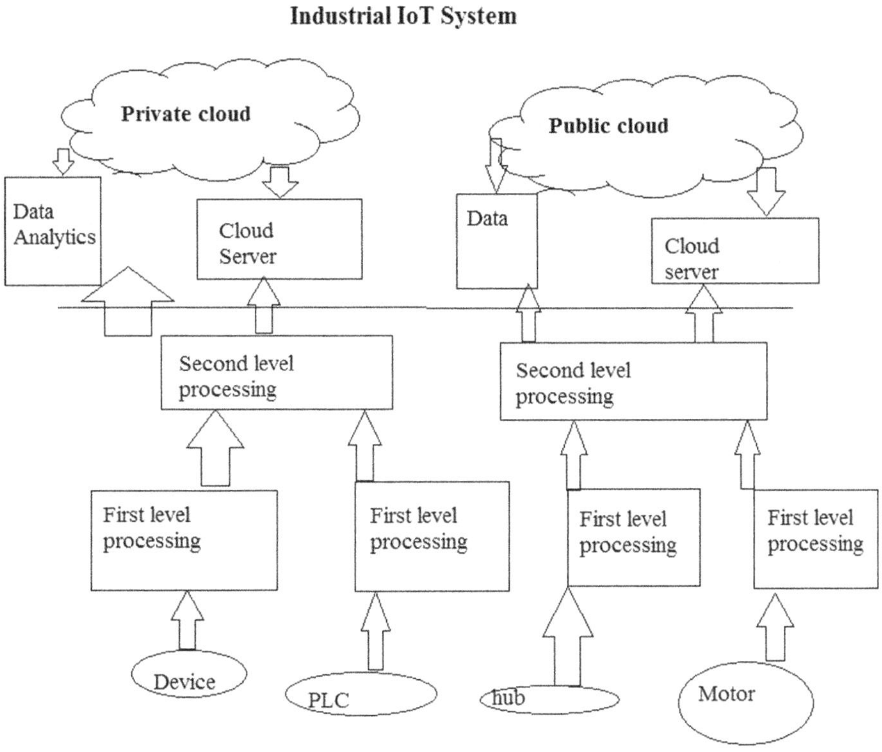

Industrial IoT System

FIGURE 9.1   Industrial Internet of Things. (From Wikipedia.)

9.1.2 EVOLUTION OF IIoT

1. **Industry 1.0:** The steam engine was invented in Industry 1.0 which evolved in 1784. The manufacturing was totally dependent on labor and it was tiresome.
2. **Industry 2.0:** The assembly line manufacturing was introduced in 1870. This reduces the labor effort of Industry 1.0 to a greater extent. Ford, the car manufacturing company, produces maximum products using Industry 2.0, which was based on the conveyor belt mechanism.
3. **Industry 3.0:** The development was mainly based on electronic technology and industrial robotics developed in 1969. Industrial robotics was involved to simplify the work, thereby it automates and increases the amount of production.
4. **Industry 4.0:** The enterprise was entirely connected through the internet in Industry 4.0, which was developed in 2010. IIoT connected with the internet has maximum advantages such as integration of product, optimization of assets, smart monitoring, intelligent decision making, remote diagnosis, and predictive maintenance.

9.1.3 ADVANTAGES OF INDUSTRIAL IoT

The advantages of IIoT are listed as follows:

1. Proactive and predictive maintenance

   Predictive maintenance for Industry 4.0 is a method of preventing asset failure by analyzing production data to identify patterns and predict issues before they happen. Organizations are implementing predictive maintenance analytics in a range of ways, from targeted solutions for a single machine part to factory-wide deployments for increasing OEE throughout the production line.

2. Real-time monitoring of production

   IIoT solutions can allow for real-time availability of supply chain data. They make it easier to track products and supplies, as well as spot slowdowns and inefficiencies. By connecting plants to suppliers, all parties involved within the supply chain can trace material flow and manufacturing cycle times. These solutions can also track operational data to original equipment manufacturers and field engineers, which allow operation managers to regulate factory units remotely and take advantage of process automation and optimization. Hence these solutions make streamline everyday workflow effortlessly.

3. Error prevention

IoT applications help reduce and often even prevent errors in inventory management and production flow. Software allows companies to monitor events across the supply chain and track inventory around the globe. Users receive notifications of important changes from the set plan as well as updates. This enables visibility across many channels into inventories and provides managers with realistic measures and estimates for materials.

9.2 CYBER PHYSICAL SYSTEM TO INDUSTRIAL INTERNET OF THINGS

CPS was developed from the massive application of embedded systems [1]. CPS is considered as the core part and foundation of Industry 4.0. CPS provides maximum benefits and will exchange the industrial operations. CPS will focus on the technologies, concept, challenges, and architecture.

9.2.1 HIERARCHICAL MODEL OF CPS

Smart manufacturing is a value-creation process from design to production, logistics, and service, in which multiple subjects participate from the perspective of hierarchy, CPS, and Digital Twins (DT) are virtual replicas of physical devices that data scientist and IT professional can use to run simulations before actual devices are built and deployed. DT can be divided into three different levels according to magnitude. They are discussed as follows.

9.2.1.1 Unit Level

The unit level refers to the smallest unit participating in manufacturing activities, such as a single piece of equipment, materials, and environmental factor. Unit-level CPS and DTs can both evolve along with physical machining, assembly, and integration in the process of cyber-physical interaction. However, a unit-level DT must be formed through the modeling of the geometric shape, identity, and function information.

9.2.1.2 System Level

In system level, multiple CPS and DTs are connected together, which increase collaboration, thereby improve the efficiency. This level constitutes the collaboration of multiple unit-level systems, and thus produces a virtual system level, for example, an aircraft that contains major components, engine of aircraft to monitor the running state, etc. System-level DTs combine together to form complex DTs.

9.2.1.3 System of System Level

Multiple system-level DTs and CPS combined together forming system of system level (SoS level). This level provides various advantages such as enterprise collaboration, massive integration, commerce-level cooperation, supply chain cooperation, high-end design. SoS level acts as a basis of foundation for innovative product development and quality-level tracing. The SoS level not only reduces the design cycle, but it also greatly lowers the cost of manufacturing in terms of money and time.

9.3 SECURITY AND PRIVACY ISSUES IN INDUSTRIAL IoT

The development of information technology to greater hike results in major demand for security. Providing protection to IIoT also results in greater challenges because of its major design.

9.3.1 ATTACKS ON IIoT

From the research literature it has been identified that the IIoT systems were prone to major cyber attacks. One of the recent attacks that found was Slammer worm. This Slammer worm damaged the monitoring system of nuclear power plant in the United States. The second attack was of a worm that infected the dispatching and signal control of the transportation system in the United States.

9.3.2 IIoT ATTACK SURFACES

IIoT has various attack surfaces, for example, the processor and the memory of electronic systems, monitoring systems, actuators and sensors, software of embedded systems, etc. IIoT systems are subject to hardware attacks, physical attacks, reverse engineering attacks, and side-channel attacks. Among these software attacks was major happened in recent days, which was due to viruses, worms, Trojan horses. The protocols used in the IIoT manufacturing among communication systems are prone to denial of service (DoS) attacks, protocol attacks, man-in-the-middle attack. The labors involved in operating IIoT machinery are prone to a phishing attack, shoulder surfing, social engineering, and social attacks.

9.3.3 IIoT Security Goals

- Protection against DoS attacks which prevent the CPS systems.
- Availability which ensures higher production.
- Periodical system monitoring that helps to prevent against damage or harm.
- Prevention against malicious attacks and viruses in IIoT systems.
- Ensuring authenticity and integrity of IIoT systems communication.

9.4 IoT CHALLENGES IN INDUSTRIAL AUTOMATION

Industrial automation spreads over various control systems such as digital control, motion control, and protection systems, etc. Various domains that have industrial automation are as follows: Batch process in which the product is developed stage by stage, for example, beverages manufacturing, food manufacturing, and pharmaceutical industries. Discrete production in which the manufacturing is done as a separate unit and package requires a separate floor.

9.4.1 Training and Determinism in Industrial Automation

Automation system requires stringent temporal behavior, maximized response, and accuracy. The major concern lies in automation systems are latency besides the throughput of the system. The best example of IoT in industrial automation is industrial robots in the welding applications where the seam direction and position are identified by the sensors. The signals obtained from the sensors have to be processed quickly such that the robot controller will change its direction quickly. The accuracy of the industrial robots is determined by their services involved in timing. Similarly, determinism of industrial robots is achieved through proper, timely operations.

9.4.2 Availability and Reliability of Industrial Automation

- MTBF (Mean time between failure): Defines the likelihood where the device, component, and system prone to fail.
- MTTR (Mean time to repair): Defines the time taken for component or system to repair after it gets fail.
- PFOD (Probability to fail on demand): Defines the likelihood where the system, component, or device that able to function properly after it was intended not.

Also, industrial automation plays a vital role in measurement and control where the IoT devices are used for decentralization and control. The metrics greatly depends on communication that is established between the data and services.

9.4.3 Safety and Interoperability in IIoT Automation

Software or operating system (OS) is not ensuring the safety and security of IIoT. Protection system has been used to enhance the safety context, for example, IEC 61508 that combines secure software with hardware applications in a different modality. An IIoT in industries contains thousands of components such as networks, servers, and work stations. Since the components are largely integrated, the possible configuration is also large. The interoperability is achieved in the industrial automation domain by means of scale and structure of components, heterogeneous issues in the diversity of components, engineering efforts needed for integration, and operation of components.

9.4.4 Fault Tolerance in IIoT Automation

The major need in industrial automation is fault tolerance since thousands of IIoT heterogeneous components are connected together and exchanging information. The fault-tolerant systems are developed in such a way that it is adaptive to failure of services, various malicious attacks, changes in the quality of services, fault related to internet connections, and software techniques. To ensure such application, software technologies and OSs have to be separated. This separation enhances the future requirements, and combined services will be provided with high concerns. Hence resilient IoT solutions have to be deployed in automation systems with fault-tolerant systems.

9.5 ANALYZING SCALABILITY IN IIoT

Scalability is the main concern in industrial automation as it contains more number of closed loops and operations. These operations require much lesser cycle time, for example, between 1s and 10s for high scalability. The automation system has to use high-frequency datasets for frequent analysis of performance. Some of the important aspects of scalability are (1)Scalable device must be sustainable in terms of its naming and addressing,(2)Adding new components or devices to existing network should not reduce the performance of an already existing system,(3)Data has to be transmitted between communication devices in higher rate with lesser cycle time.

9.5.1 Scalable Collaboration in IIoT

In industrial automated systems, all the controllers and components are connected through networks based on Ethernet. These components can share their data in larger hierarchy, and connection has to be vertical silos because they are point-to-point. Explicit engineering activities have to be performed in both horizontal and vertical directions of components for proper communication and collaboration.

9.5.2 Scalability with Security in Automation

Functional safety represents the overall critical safety of the automation system. Examples are oil and safety systems, nuclear power plants; all rely on functional safety. The concern about functional safety relies on both software and hardware. As most of the products and systems are interconnected, functional safety ensures the proper services of devices and paramount. In addition to functional safety, data integrity is needed as the components are interconnected, and data are collected from sensors, actuators located in remote places. Therefore, there is a need for IoT devices to provide enhanced security with a less computational cost. Some of the automation systems provide high scalability in microseconds with all upgradability. Safety integrity control (SIC) can be integrated with the automation system to prevent accidental control actions.

9.6 MANAGING INDUSTRIAL IoT CHALLENGES

This section describes about the possible potential solutions for IIoT challenges.

9.6.1 DATA SCALABILITY AND LATENCY BASED ON LOCALIZATION OF COMPUTATION

Edge computing processes the application logic in which data are transferred from corporate premises to the edge network. This kind of processing reduces network latency and increases responses to end-users. This is more suitable in real-time applications, such as distributed systems, the network that requires reduced latency, and improved quality of service (QoS). Similarly, Fog computing supports internet of everything (IOE) where it gives predictable latency. Fog computing finds its application in industrial automation, robotics, transportation, IoT sensors and actuators, big data analytics, etc. In addition to Fog and Edge computing, computational offloading is another way of reducing network latency, network bandwidth, storage capacity. This method increases processor performance in industrial automation. The service orchestration and service choreography describes an architecture that provides efficient service integration based on local orchestration of web-enabled services in distributed form. It also provides peer-to-peer (P2P) architecture for IoT networks that aims to provide automated services. The major application of this architecture is automated resource discovery mechanism without human intervention and configuration.

9.6.2 BALANCING FAULT TOLERANCE

The most common approach for managing fault tolerance is redundancy in systems. The redundancy in IIoT includes duplicate controllers, servers, IO devices. The fault-tolerant algorithm is designed such that it is running continuously in asynchronous manner. This enables the guarantee of fault-tolerant when failures occur in the system. Specialized controllers are also available that provide built in redundancy using dual-processor modules. Local fall back solutions are needed in industrial automation that has reliable and deterministic data analysis and data transfer in real time. This mechanism removes the single point of failure in automation system. DoS attack can be reduced and service availability can be increased using redundant nature. Hardware faults can be addressed using fault prediction mechanisms that analyze and process the information in preventive maintenance.

9.6.3 MIXED CRITICALITY BASED ON SYSTEM PARTITIONING

Virtualization is the emerging technique that finds its application in embedded real-time system and industrial systems. Virtual machine monitors the usage of physical resources, such as RAM, CPU, network interfaces. Each virtual machine has its own OS and hypervisor, sometimes manages the physical resources. Virtualization enables the integration of different OS in multi-core platform. This portioning results in a higher degree of independence. It also maintains the running of legacy software in OS. System portioning also results in larger security updates, and it ensures high stability in real-time OS. Most of the software in industrial automation is single core. To partition the single core to multicore spatial, temporal segregation is required. Criticality in industrial automation is guaranteed with partition model with virtualization.

9.6.4 BALANCING SECURE AND SCALABLE REAL-TIME COLLABORATION

Scalable collaboration needs an automatic configuration of connected devices without manual intervention. Zero-configuration is the emerging network technology that ensures faster communication between the network devices and ease of use. It is based on three technologies such as (a) Link local addressing (LLA): assigning network devices with numeric address,(b) Multicast DNS name resolution (MDNR): automatic hostname distribution, and (c) DNS service discovery (DSD): automatic discovery of services such as cameras, gateways, printers, speakers, etc. Many communication technologies are available that connect heterogeneous devices and its applications. Some of the protocols are Advanced Message Queuing Protocol (AMQP), Message Queue Telemetry Transport (MQTT) and data distribution service (DDS).

9.7 TECHNOLOGIES INVOLVED IN PROTECTING INDUSTRIAL IoT

Protecting the IIoT involves the discovery of new threats and vulnerabilities. IIoT becomes more manageable if all the software, policy, and updates are up to date. The deployed version of automation system should be carefully controlled, managed, and configured. Periodic compliance reports about security is advisable and mandatory. Similarly, networks and its endpoint configurations should be periodically monitored for any deviations.

9.7.1 SECURE OPERATIONAL MANAGEMENT

It defines protecting the operational management process and functions, such as to ensure the integrity and confidentiality. Operational management must interact with the operational monitoring. Operational and security events must be separated as they can be used to security gaps in the network, communication, and interfaces. Application Programming Interface (API) is used to segregate the security and operational aspects. Three most common APIs in security management are receive policy, gather logs and communication events, and gather endpoint properties.

9.7.2 Security Policy Management

There are three types of policy management:

1. Machine policy: Consists of digital document that includes settings for technical security controls.
2. Regulatory policy: High-level policy that separates good behavior from bad.
3. Organizational policy: It documents behaviors such as technical and nontechnical.

9.7.3 Security Model Change Control

Security model changes should occur for endpoint depending on the state of the life cycle. The different stages of the life cycle in security management are specified as follows:

1. Commissioning: It provides a temporary identity for all endpoints. Literally, the system builder and the component builder must commission the endpoint.
2. Provisioning: It replaces the organization identity with the trust root identity. It will set all the endpoints to normal use.
3. Commission usage: Optimum security is applied to the endpoint during this stage. Based on the security event that occurred, endpoint may transfer to the alert state that enhances the security and minimizes the operational functionality.
4. Endpoint decommissioning: In this stage, lifecycle of endpoint gets terminated. This decommissioning can be reused and re provisioned for some other purpose.

9.7.4 Security Considerations for Monitoring

Supply chain is a special consideration for security monitoring as it requires the monitoring activities for producing IIoT components. The security and integrity are also ensured in this case of using supply chain. The monitoring state must always be consistent with a network, security policies, and endpoint behavior. Some most commonly used considerations are brownfield considerations, security and privacy considerations, data logging considerations, etc.

9.8 APPLICATIONS OF INDUSTRIAL IoT

IIoT is rapidly eminent sector for maximum share in global IoT spending [2]. Manufactures and industrialist monitor the automation of complex process in manufacturing. Some of the major applications of IIoT are specified as follows:

9.8.1 Digital Factory

Using IoT the data can be easily transmitted to the remote manufacturers and field engineers. This enables the owners to remotely monitor the factory units and their working conditions. Digital factory provides automotive processing and optimization of working. In addition to this, owners can easily identify the key result areas (KRA) using this digitalization.

9.8.2 Managing Facility

Sensors and actuators in IoT help to provide the maintenance alerts in industrial automation. Sensors can provide data based on vibration and temperature changes; it will continuously monitor the working condition of automotive system which sends alerts to owners if any deviation occurs in the prescribed format of working. This gives the advantages for owners in the way of cost reduction, energy conservation, reducing machine downtime, and increases operational efficiency.

9.8.3 Production Flow Monitoring

IIoT in automation helps to monitor the production from starting of refining process to packaging the products. Thus, the complete monitoring of production provides scope for manufacturers in recommending adjustments of operation and operational efficiency. Flow monitoring also eliminates the lag in product failure.

9.8.4 Inventory Management

IoT can also give way for supply chain monitoring; hence the inventory can be traced globally. Monitoring happened on line level, and maintenance alerts will be given to owners on any deviations. The providers will able to get cross channel visibility on inventories and realistic estimates of products.

9.8.5 Plant Safety and Security

IIoT combined with big data analytics, enables the overall working and optimization in plant. The key performance indicators (KPI) such as short-term absences, long-term absences, near illness, number of misses, health safety, damages to machinery, loss on damages can be easily monitored. Thus, the effective monitoring results in enhanced safety [3].

9.9 DESIGN PATTERNS IN INDUSTRIAL IoT

For all complex designs, systematic approach is needed, which is specified by system engineers, operational analysis, and system designers. Design patterns are used to identify reusable components within the system and provide solutions for recurring problems. The design patterns should satisfy the following requirements:

1. It should be abstract to define computing and communication protocols.
2. It should be composable for multiple patterns to combine.
3. It should be recognizable for all components in IIoT settings.
4. It should be data centric for a better understanding of resources.

Based on the above requirements, five design patterns have evolved in IIoT. They are:

9.9.1 CLOSED LOOP

The main advantage of this pattern is it extracts information from process automation for delayed transmission and stores them in cloud for future analysis. This pattern periodically reads the data from the sensor and will produce control signals based on the difference between input and output. Based on the control produced, actuators of IIoT will take action. The closed-loop pattern establishes network communication between sensors and actuators. It is also used for establishing data collection and communication off the critical path.

9.9.2 OPEN LOOP

This design pattern provides commands to the system unilaterally; it is useful in the low-cost applications. The open-loop patterns recognize the process automation system when there is a failure in the communication link and error in sensors. The logging action of IIoT system provides insight into failure rate of primary controllers. The closed and open loops are given in Figures 9.1 and 9.2.

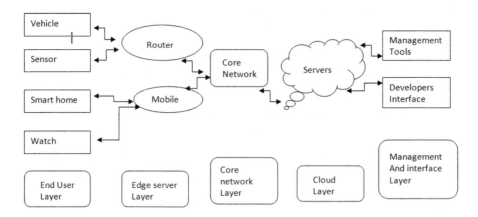

FIGURE 9.2   Scalability in IIoT.

9.10 CORRELATION WITH INDUSTRIAL IoT REFERENCE ARCHITECTURE

IIoT reference architecture provides common framework for high level abstraction. It provides identification and comprehension of the most important issues. IIoT reference architecture is a standard-based open architecture for industrial consortium. The high level abstraction provides broad industrial capability. The ISO/IEC/IEEE 42010:2011 system and software engineering architecture description provides standard conventions and practices. It provides an ontology for the description of architectures and provides system architect to express the architecture. A viewpoint comprises the framing and analysis of specific system concerns. The stakeholders and constructs of viewpoints together are called as frames of architecture. The architecture view, along with architecture models, is known as architecture representation. The ISO/IEC/IEEE architecture specifications have generally been considered by means of concerns, view points, and stake holders.

9.10.1 INDUSTRIAL INTERNET VIEW POINTS

The industrial IoT reference architecture (IIRA) viewpoints can be defined by analyzing various use cases of IoT. The four viewpoints are (a) business viewpoint, (b) usage viewpoint, (c) functional viewpoint, and (d) implementation viewpoint.

9.10.1.1 Business Viewpoint

Business viewpoint concerns about the identification of stakeholders and their vision, objectives, values in IIoT system. It further explains about stated objectives through its mapping to all fundamental capabilities. These concerns are business-oriented, and it is of major interest to decision-makers, product managers, and system engineers.

9.10.1.2 Usage Viewpoint

The usage viewpoint implies the expected system usage. It involves the sequence of activities in human or logical users. The intended capability of usage viewpoint is achieving all fundamental system capabilities. The stakeholders of usage viewpoint are system engineers, project managers, and other stakeholders.

9.10.1.3 Functional Viewpoint

This viewpoint concerns about functional components in IIoT system, their structure, and inter relation. It also concerns about interfaces that exist in the system, relation, and interaction of the system with external elements in the environment. It also supports the activities and usages of overall systems. It is of major interest to system architect, component architect, developers, integrators, and system operators. Figure 9.3 depict the industrial internet references architecture.

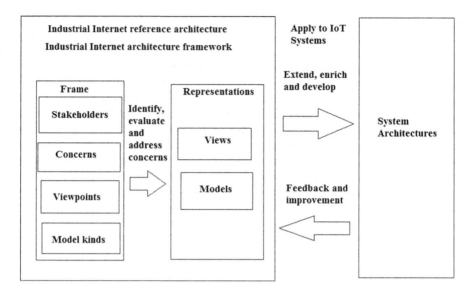

FIGURE 9.3   Industrial internet reference architecture.

9.11 IIoT IN CYBER MANUFACTURING SYSTEMS

Cyber manufacturing system represents a translation of data from strongly connected components into perspectives and predictive operations to achieve greater performance. Cyber manufacturing system (CMS) and IIoT systems are not separate technologies instead, they are interconnected. The main challenge with CMS and IIoT is communication among interconnected devices [4]. Efficient communication is necessary for real-time delivery of information with robustness and all quality of services. The interconnected system also needs a greater level of abstraction and self-management [6–8].

9.11.1 Modeling of CMS

Modeling refers to controlled approach facilitating the composition of whole system parts. The research work for smart manufacturing system design and analysis (SMSDA) provides a resilient framework in four project investigations. They are:

- Modeling methodologies for analysis.
- Methodologies for predictive analysis.
- Performance measurement.
- Manufacturing based on service and service composition.

9.11.1.1 Modeling of CMS for Analysis

Manufacturing of the new model should ensure feasible and optimal system results. The developing model does not follow the precise methodology and predictive methods should be used for systematic analysis. This type of manufacturing gradually increases the cost and time of making actionable recommendations. The manufacturing model also ensures the synthesizing of elements in functional, equation based, and domain specific modeling methods. Modeling of CMS, which deals with optimization problems, becomes integral part of smart cyber manufacturing system components.

9.11.1.2 Predictive Analysis for CMS

The principal foundation of CMS is a predictive analysis of data, whereas the existing predictive analysis is mainly depend on proprietary models. To overcome these challenges, the development should contain:

- Solution for fusion, data capture, filtering, and dimension reduction.
- Measurement-based methodology for uncertainty, traceability, security, validation, and data provenance.
- Predictive Markup Modeling Language (PMML) is designed to extent manufacturing applications.
- The prototype system should include predictive analytics and domain-specific languages.

9.11.1.3 Performance Measurement for CMS

Performance measurement needed for smart manufacturing of CMS requires agility, productivity, quality, and sustainability. These systems integrate software applications for optimizing a variety of performance metrics and optimize the high-quality measurements. Sustainability in manufacturing is defined as developing manufactured products that conserve energy, which is non-polluting, economically safe for employees, consumers, and natural resources involved. As productivity and agility of CMS increases, QoS and sustainability also increase. Although various opportunities available, there are still challenges available in identifying the opportunities in increasing the performance metrics.

9.11.1.4 Service-Based Manufacturing and Consumption for CMS

The rapid evolution and traditional CMS is done by the integration of software subsystems through the proprietary exchange of data. Cloud computing is one of the emerging technologies for service-based manufacturing and consumption. The Smart Manufacturing Operations and Planning Control (SMOPC) focuses on the integration of technology for service-oriented. The SMOPC systems enable the rapid manufacturing with integration and collaboration for optimizing the system performance. The advantages of SMOPC systems are:

- Usage of contextual information for life cycle management and precise cataloging.
- Development of system with evolution of core messaging standard.
- Expanding context meta model for various association types.

9.12 ARTIFICIAL INTELLIGENCEINTEGRATION WITH IIoT

IIoT devices generate a huge amount of data that can be used with artificial intelligence (AI) systems. Some of which are facial image recognition used in a border control system, autonomous cars generate data about surrounding using sensors. Existing IoT platform uses many interfaces that gather information from various devices that utilize all the data into machine learning or AI devices. The following three examples show how logistically AI with IoT can be used to solve complex problems [5]. They are EI City Brain, Tesla's Auto Pilot, and Classroom monitoring system.

9.12.1 EI CITY BRAIN

EI City Brain developed by Alibaba Cloud is a more complex AI solution that utilizes urban public resources. It is implemented in most countries which reduces traffic by 15%. The solutions can also help to identify road accidents, illegal parking, and support ambulances to reach the target faster by changing the traffic light. It is an emerging software system which utilizes traffic light cameras all over the city and is based on output from machine learning, which determines how to change the traffic light.

9.12.2 TESLA'S AUTOPILOT

This autopilot system in most countries uses GPS, sensors, cameras, and forward looking radars utilizes data that can be coupled to neural network architecture. It works as self-enclosed systems which collect data from sensors, and neural network model can be used to determine what should be the next change in the movement of the car [6].

9.12.3 CLASSROOM MONITORING SYSTEM

The classroom monitoring system can now be implemented in many developing countries. The classroom cameras scan the room once in 30 seconds, and the algorithm will determine the student's emotion as happy, sad, bored, etc. It also scans the behavior of students, for example, reading, writing, and raising hand, etc. In this monitoring system, cameras gather data, and image recognition is performed to predict the result.

REFERENCES

1. Sabina Jeschke, Christian Brecher, Houbing Song, Danda B. Rawat. 2017. Industrial Internet of Things cyber manufacturing systems. *Springer Series in Wireless Technology.* http://www.springer.com/series/14020
2. In lee, Kyoochun Lee. 2015. The Internet of things (IoT): Applications, investments and challenges for enterprises. *Business horizon*, Vol. 8, Issue 4, pp: 431–440.

3. Mauro Conti, Ali Dehghantanha, Katrin Franke, Steve Watson. 2018. Internet of things security and forensics: Challenges and opportunities. *Future generation computer systems*, Vol. 78, pp: 544–546.
4. Muna Al-Hawawreh, Nour Moustafa, Elena Sitnikova. 2018. Identification of malicious activities in industrial internet of things based on deep learning models. *Journal of information security and applications,* Vol. 41, pp: 1–11.
5. Emiliano Sissni, Abusayeed Saifullah, Song Han. 2018. Industrial Internet of things: Challenges, opportunities and directions. *IEEE transactions on industrial Informatics*, Vol. 4, Issue 11, pp: 4724–4734.
6. K. K. Christopher, X. V. Arul, P. Karthikeyen. 2019, April. Smart toll tax automation and monitoring system using android application. In 2019 IEEE International Conference on Intelligent Techniques in Control, Optimization and Signal Processing (INCOS), IEEE, pp. 1–6.

10 Unifying Cloud Computing with Internet of Things Using Secured Protocol

S. Rakoth Kandan
Jayamukhi Institute of Technological Sciences
Warangal, Telangana, India

N. Dhanasekar
A.V.C. College of Engineering
Mayiladuthurai, Tamil Nadu, India

P. Avirajamanjula
Prist University
Thanjavur, Tamilnadu, India

CONTENTS

10.1   INTRODUCTION: BACKGROUND AND DRIVING FORCES

Internet of Things (IoT) is a network of various interconnecting applications, physical objects, devices, vehicles, and other items used by networking technologies [1]. IoT is the new revolution among the technology sector; it holds a very big difference among business purposes. McCue described that the IoT market segment holds to hit $117 billion by 2020.

IoT is used in day-to-day life applications and has created a high impact on potential users. A single private user can get access to IoT in both home and working environments [2]. In home appliances it plays an important role, i.e., e-health, assisted living, enhanced learning, etc. Similarly, IoT is also used in business appliances, i.e., transportation, industrial manufacturing, etc. [3].

IoT is a technology that is surrounded by software, sensors, and electronics that support to succeeds by connecting devices. Mobile cloud computing (MCC) is the technology that supports for supreme capability and performance of the existing structure [4]. MCC is based on cloud computing, which provides access to resources from any location and reducing the hardware devices usage. MCC is a platform which interlinks the cloud computing and smartphones to utilize the resources more effectively [5]. The term mobile cloud is mentioned in two manners (i) infrastructure-based, and (ii) ad hoc mobile cloud. It could be used for mutually IoT and video surveillance technologies and could offer developments on their roles.

10.2   LITERATURE REVIEW

This section will discuss the review of the related researches in integration with cloud computing and IoT. Several researchers analyzed and published previously, Subhashini et al. survey different security risks in a cloud environment and also focused service delivery models in a cloud computing system [6]. Hassan et al. presented a trustworthy environment to provide an exploration of the roadblocks and its solutions [7]. George et al. proposed a new platform using cloud computing for real-time applications, smart city services, and presented a framework for highly distributed data, heterogeneous, decentralized, real devices, i.e., sensors, actuators, etc. It can be controlled using distributed cloud services [8]. Fei et al. proposed the architecture, IoT and cloud computing based manufacturing service system (CMfg), the relationship among three terminologies, i.e., CMfg, cloud computing, and IoT is studied [9]. Jiehan et al. proposed the cloud things architecture that accelerates IoT applications and its development, also reviewed the integration among IoT and cloud computing and observing an IoT-enabled smart house appliances [10]. Antonio et al. discussed the challenging factors among cloud computing and IoT integration and enabling the IoT based transportation system to reduce the fuel prices, carbon dioxide emissions, traffic congestion, ways to improve street security [11]. Moataz et al. developed smart home appliances by integrating the IoT, web services with cloud computing technologies. Also, it implemented three use cases to implement in the home appliances using this method, which provides possibility and efficacy [12]. Jayavardhana et al. presented an essential vision for cloud computing execution of IoT. The different technologies and applications in IoT research areas are discussed [13]. Cloud implementation by using the Aneka method on the public and private cloud also discussed. Aazam et al. presented the integration importance of IoT with cloud computing because the resource consumption and storage volume have the privilege, even though to make it more useful, this type of integration is suggestible [14]. Mohammad Aazam et al. presented a new model named Cloud of Things (CoTs) for the integration of cloud computing with IoT, In further to increase the valuable services, this integration is essential [15]. Alessio Botta et al. surveyed the integration between cloud computing and IoT, we mention it as CoT model [16]. Mohammad Aazam et al. proposed the key challenging task in CoT and smart way based communication. The main advantage of CoT is performing the right tasks, whether not possible, using IoT [17]. Manuel et al. survey in integration components, data analytics, and integration mechanism between IoT and cloud computing are discussed [18]. Mohammad Alsmirat et al. focused on segmentation-based algorithms particularly the Fuzzy C-Means algorithm, which provides accurate predictions [19]. Gupta et al. focused on face recognition methods in biometric security mechanisms [20]. Zhiyong Zhang et al. surveyed the effective mechanisms for security in social media tools and presented a hierarchical method for evaluation and measuring [21].

10.3 IoT MODEL

The "Internet of Things (IoT)" developed initially in "Future of the Internet and Ubiquitous Computing" and was perceived by a British scientist "Kevin Ashton" [22]. IoT is a system that can able to connect the world using the internet-enabled sensor objects. IoT connects with various devices using different protocols [23]. The IoT offers a flexible and suitable agenda for peoples to interrelate with the situation around us. Basically, IoT is using the interconnection with small objects, these methods used to identify the environmental situation around them, after sensing the situation sensors share this information using the internet for further processing [24].

10.3.1 HISTORICAL METHOD

Since IoT is popular globally and providing services reliably, Intel specifies IoT as fixed services on the internet. Today all the home appliances have the ability to operate and control using the internet. IoT architecture is basically divided into four layers: Application, Middleware, Physical, Network/Transport, Perception/Physical Layers. Each layer functions as mentioned in Table 10.1.

TABLE 10.1

IoT Layers and Functions

S. No	Layers	Functions
1.	Application	Smart city, Smart healthcare, Smart home, Smart agriculture
2.	Middleware	Service management, Data storage, Service composition
3.	Network/Transport	LTE, 3G, GPRS, GSM, Wi-Fi, PSTN/ISDN3
4.	Perception/Physical	Sensors, RFID's, Bluetooth, GPS
1.	Application	Smart city, Smart healthcare, Smart home, Smart agriculture
2.	Middleware	Service management, Data storage, Service composition
3.	Network/Transport	LTE, 3G, GPRS, GSM, Wi-Fi, PSTN/ISDN3
4.	Perception/Physical	Sensors, RFID's, Bluetooth, GPS

10.4 CLOUD COMPUTING MODEL

Cloud computing provides a platform with on-demand service for database, data analytics, computational power, and software. It involves data storing on a remote location server and perform the computation process using virtual machines, which reduce the managerial task on guest users [25]. The cloud platform favorably provides services like economical, reliable, elasticity, scalable, security, and performance. Cloud computing deployment models and various services in three distinct levels, Infrastructure, Software and Platform, are mentioned in Tables 10.2 and 10.3.

TABLE 10.2
Cloud Computing Deployment Models

S. No	Types of Cloud	Functions
1.	Public cloud	Elasticity, utility pricing, available for all,
2.	Private cloud	Outsource or own, private use, total control
3.	Community cloud	Meets shared concerns, Several stakeholders,

10.5 INTEGRATING INTERNET OF THINGS AND CLOUD COMPUTING

Figure 10.1 depicts the cloud IoT model. Internet-based computation has rapidly developed and increased the number of interconnection with each other. Data storage in IoT is a difficult task; it does not support data stored locally, previously sensors send the data to supercomputers, which support resources. This method had some inconveniences, such as time-consuming, high costs. Another method is distributed computing where nodes are used for data storage and processing, this method had some inconveniences that replacement of IoT nodes in failure condition is difficult and it takes more cost. In recent years, IoT devices available at low cost and provides more computing power. The IoT devices are used for sensing the infrastructure, data capturing, and transmission purposes using internet services where the cloud environment provides these services in a systematic and flexible computing resource [26]. The two divergent technical platforms are cloud computing and IoT, in these, different cloud layers are connected to IoT sensors, objects, and other services. The following four are considered as the motivating features for the cloud IoT model: Storage, Computing Capabilities, Communication, and New capabilities and Paradigms.

TABLE 10.3
Cloud Service Models

S. No	Services	Functions
1.	SaaS—Software as a service	E-mail, Google apps, CRM, virtual desktop
2.	PaaS—Platform as a service	Runtime environment, database, web server, tools
3.	IaaS—Infrastructure as a service	Virtual machine, Servers, Network storage

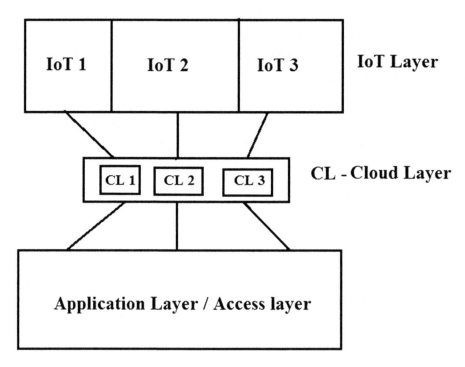

FIGURE 10.1   Cloud IoT model.

10.5.1   CLOUD COMPUTING—IOT APPLICATIONS PLATFORM

Figure 10.2 will show that integrating cloud and IoT created more opportunities to utilize the distributed computing environment [27]. There are three different possibilities in the cloud IoT environment to achieve this, machine-to-machine (M2M) sensor nodes generate an interrupt when some expected events occur. Human-to-machine (H2M) is human communicating with a machine using the voice recognition technique. Machine-to-Human (M2H) is understanding the human features by using biometric devices.

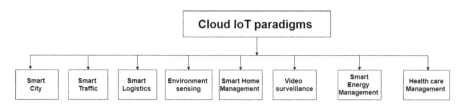

FIGURE 10.2   Cloud computing—IoT applications platform.

10.6 CLOUD IoT SECURITY RISK AND PROBLEMS

The cloud IoT model interconnects the devices, provides services like smart city, smart traffic, smart logistics, environment sensing, smart home management, video surveillance, smart energy management, healthcare management [28]. The cloud environment provides an efficient platform for different services by sharing the computational power, storage, data analytics, and mining abilities. The cloud IoT model is proved to be beneficial in upgrading day-to-day life cycle [29]. In case suppose any susceptibility activity in the cloud IoT network, all the processes can be extracted, and the diagnostic information can be harmed by hackers. The integration model will further worsen the situation, and such activities will bury issues and susceptibility. The loopholes in the cloud IoT model will overcome the real benefits, and also it does not possible to periodically change the devices like sensors in the IoT model [30, 31]. Hence the model should be robust and constant workflow for a minimum period without replacement and maintenance.

10.6.1 DIFFERENT PRESERVATION FEATURES AND GOALS

Table 10.4 shows the cloud IoT preservation goals. In cloud IoT model, data transmits between the devices and user to reach the exact destination. To achieve the safe communication infrastructure, few parameters are considered as essential factors, i.e., privacy, authentication, and controlling [32].

10.7 RELATED RESEARCH ON CLOUD IoT SECURITY

The integration of the cloud IoT model needs a security measure in both cloud and IoT environments. To achieve this goal, as mentioned in Table 10.4, data privacy and integrity are the main issues in the cloud IoT model. Data confidentiality also plays a key role in this model, some previous cases cryptographic techniques, protocols, privacy, and identity management are used for integration purposes [33].

Techniques like asymmetric mutual authentication technique, secure hash algorithm, and elliptic curve cryptography are used for authentication purposes, feature selection, platform and node sinking, etc. [34, 35]. Another technique used based on the request and reply method by using a key matrix method is called a one-time cipher. Encryption and decryption technique used in this communication. To secure

TABLE 10.4

Cloud IoT Preservation Goals

S. No	Security Goals
1.	Availability
2.	Integrity
3.	Privacy and identity
4.	Confidentiality and authentication

the IoT nodes from network-based attacks, a framework is used for integration technique based on authentication and authorization. The proposed technique is using the security-based protocol, and the obtained results are satisfying.

The perception layer contains the sensors and RFID's elements which have controlled computational capability. In such a way, the use of complex cryptographic procedures is not possible, to avoid this inconvenience, the XOR-based encryption technique is used [36]. This rule avoids the complex issues in the cloud IoT model and safeguards the authentication procedure in an RFID system.

10.7.1 Possible Protection Methods

The cloud IoT model should focus on both security and IoT layers and the cloud model, to avoid the illegal node access in the perception layer, authentication is mandatory [37]. The encryption techniques need to confirm data reliability. In such a way, "elliptic curve cryptography" (ECC) is applied with key generation and exchanging procedure [32]. To increase the life span, energy-saving methodologies must use in the cloud IoT model, to achieve the optimal solution in energy savings renewable sources like solar, wind, and air must utilize. To reduce the physical damage in sensor nodes, the supervisor should check and analyses from the remote location. To avoid attacks such as denial of service, distributed denial of service, man-in-the-middle the firewalls policies, and filtering techniques. It is more encouraged that end-to-end encryption techniques should use for implementation; it will help to maintain the constant integrity among data [33]. To protect the application layer from attacks, such as "cross-site scripting" and various threats, the scripting and coding techniques should be very effective. To safeguard the data in cloud environment, protocols need to be formed and implemented. The encryption should be properly carried out for preventing data leakage and other threats. To identify the anonymity activity in the cloud IoT model, the protocols will help to store the data in a secure file using various operations. The cloud IoT model integrates two different platforms, security plays a key role in both techniques, and it achieves with different protocols based on security levels.

10.8 CLOUD IoT INTEGRATION CHALLENGES AND PARAMETERS

It is wholly visible that integration of the cloud IoT model will create discernible changes in our day-to-day life, anyway, the integration also creates some severe issues that must be taken care of by the researchers in particular security aspects [38].

10.8.1 Security Issues

Both cloud and IoT are dispersed in nature. To improve the security, privacy infrastructure is a key role in its integration. To preserve the critical data in the cloud environment, the design of optimal measures are needed along with data integrity, data availability, and authenticity [27]. In a cloud environment, TPA is using in some places for data sharing between users, and cloud which may propagate for illegal activities. So, the framework needs to be created for heterogeneous communication among IoT nodes as well as safeguard the data in a cloud platform.

10.8.2 DIFFERENT STANDARDS AND PROTOCOLS

Since there is no standardized IoT architecture is available, different rules need to design for communication and exchange information with each other. Even though the nodes belong to a homogenous network, the rules are like to be heterogeneous, like Zigbee, IEEE 802.15.4, CoAP, etc. There are some chances for data collection gateway would not sustenance the rules and prominent to unsuitability issues. If cloud integrates with IoT, there are some changes to the severe problem occurring; hence the research scientist needs to improve the rules and scalable platforms, so the unified integration service can achieve.

10.8.3 EFFECTIVE POWER UTILIZING

The universal communication between cloud and IoT generating more data, so the battery becomes drain because of power-constrained in the sensor node. It will be severe in the case of visual data processing. Since the sensor node is back up with battery power, it is not possible to change in a frequent period. The solution to this problem may be an alternate way of resources like air, wind, solar power [39]. Another suggestive method for this problem is to assign sleep mode in a regular period for the sensor nodes.

10.8.4 DELAY AND LIMITED BANDWIDTH

The cloud platform contains the huge resources and different services; however by using this service, there is no guarantee that latency and delay will reduce. To get an optimal solution, bandwidth should be high for data transmission. By placing the fog computing in between the cloud and IoT, the delay can be reduced [27, 40].

10.8.5 QUALITY OF SERVICE

The performance of internet work can be identified using the parameter is quality of service (QoS). The huge volume of data is generated and converted for cloud IoT model usage, the main objective is to maintain the QoS in a particular platform is the most important thing. The request from the client can be considered under demand efficient management in the cloud platform may be delay-sensitive. To avoid packet loss in QoS, the techniques need to improve, and data packets must be the perfect solution.

10.9 DISCUSSION AND CONCLUSION

Cloud and IoT have become key factors in our day-to-day life for accessing different applications. The integration of cloud and IoT is essential for good infrastructure, high availability, good performance, etc. The cloud environment provides a well-organized model and avoids several issues that occur in IoT systems. Earlier, most of the people concentrated separately on either cloud or IoT. To overcome the new research era, this section conceded a deep review of the various research papers and presented a global vision for cloud IoT model integration. The security issues were highlighted, which affects the IoT architecture, including the threats in the cloud environment. Also various measures for potential mitigation were discussed and identified. The main aim is to rectify the heterogeneity issue among cloud and IoT. Finally, the chapter discussed the challenging factors in the cloud IoT integration model. Based on the literature review, it is clear that more steps may be necessary to achieve perfect and faultless merging of cloud IoT integration. Many works need to be carried out on security, designing, and encryption procedures so the devices and nodes can properly utilize the access level IoT network and cloud environment. Data privacy in the cloud IoT model needs to increase, so the integration of a particular system will be maintained. The designed rules must consider minimizing the usage of power resources. To save the energy resources to exploit any alternate form of energy generation, the efficiency of the system can be increased by given sleep mode in a regular tie period for the sensor nodes in the IoT network. The term called fog computing is sensitive and will make sure that less latency, transmission delay in both cloud and IoT. Lastly, QoS in data transmission was considered, which deals with IPv6 features like flow label, and different traffic classes were suggested.

REFERENCES

1. T. J. McCue, *Billion market for internet of things in healthcare by 2020*, 2015. Retrieved April 22, 2015, http://www.forbes.com/sites/tjmccue/2015/04/22/117.
2. Luigi Atzori, et al., The Internet of things: A survey, Comput. Netw. 54 (2010), 2787–2805.
3. Sandip Roy, et al., A fog-based DSS model for driving rule violation monitoring framework on the Internet of things, Int. J. Adv. Sci. Technol. 82 (2015), 23–32.
4. Y. Kryftis, G. Mastorakis, C. Mavromoustakis, et al., Efficient entertainment services provision over a novel network architecture, IEEE Wireless Commun. Mag. 23 (1) (2016), 14–21.
5. S. Fremdt, R. Beck, S. Weber, Does cloud computing matter? An analysis of the cloud model software-as-a-service and its impact on operational agility, in: 46th Hawaii International Conference on System Sciences, pp. 1025–1034, 2013.
6. S. Subashini, V. Kavitha, A survey on security issues in service delivery models of cloud computing, J. Netw. Comput. Appl. 1 (34) (2010) 1–11.
7. Hassan Takabi, James B. D. Joshi, Security and privacy challenges in cloud computing environments, IEEE Security & Privacy. Vol. 8 (2010), 24–31.
8. George Suciu, et al., Smart cities built on resilient cloud computing and secure Internet of things, in: 2013 19th International Conference on Control Systems and Computer Science, Bucharest, 2013.
9. Fei Tao, et al., CCIoT-CMfg: Cloud computing and Internet of things-based cloud manufacturing service system, IEEE Trans. Ind. Inform. 2 (10) (2014) 1435–1442.
10. Jiehan Zhou, et al., Cloud Things: A common architecture for integrating the Internet of things with cloud computing, in: Huazhong University of Science and Technology, Wuhan, 2013.

11. Juan Antonio Guerrero Ibáñez, et al., Integration challenges of intelligent transportation systems with connected vehicle, cloud computing, and Internet of things technologies, IEEE Wirel. Commun. Vol 6 (2015), 122–128.
12. Moataz Soliman, et al., Smart Home: Integrating Internet of things with web services and cloud computing, in: 2013 IEEE International Conference on Cloud Computing Technology and Science, Oulu, 2013.
13. Jayavardhana Gubbi, et al., Internet of things (IoT): A vision, architectural elements, and future directions, Future Gener. Comput. Syst. (2013) 1645–1660.
14. Mohammad Aazam, et al., Cloud of things: Integrating Internet of things and cloud computing and the issues involved, in: Proceedings of 2014 11th International Bhurban Conference on Applied Sciences & Technology, IBCAST, Islamabad, 2014.
15. Mohammad Aazam, et al., Cloud of Things: Integration of IoT with Cloud Computing, Springer International Publishing, 2016, pp. 77–94.
16. Alessio Botta, et al., Integration of cloud computing and Internet of things: A survey, J. Future Gener. Comput. Syst. (2015) 1–54. 14/09/.
17. Mohammad Aazam, et al., Smart gateway based communication for cloud of things, in: 2014 IEEE 9th International Conference on Intelligent Sensors, Sensor Networks and Information Processing, ISSNIP, Symposium on Public Internet of Things, Singapore, 2014.
18. Manuel Díaz, et al., State-of-the-art, challenges, and open issues in the integration of Internet of things and cloud computing, J. Netw. Comput. Appl. (2015) 99–117. 25/09/.
19. Mohammad Alsmirat, Yaser Jararweh, Mahmoud Al-Ayyoub, Mohammed A. Shehab, B. B. Gupta, Accelerating compute intensive medical imaging segmentation algorithms using GPUs, in: MTA, Springer, 2016.
20. B.B. Gupta, D.P. Agrawal, Shingo Yamaguchi, Handbook of Research on Modern Cryptographic Solutions for Computer and Cyber Security, IGI Global Publisher, USA, 2016.
21. Zhiyong Zhang, Brij B. Gupta, Social media security and trustworthiness: Overview and new direction, Future Gener. Comput. Syst. (2016) Elsevier.
22. Miao Wu, Ting-Jie Lu, Fei-Yang Ling, Jing Sun, and Hui-Ying Du, "Research on the architecture of Internet of things." In: 2010 3rd International Conference on Advanced Computer Theory and Engineering (ICACTE), vol. 5, p. V5–484. IEEE, Aug 2010
23. Matthias Kovatsch, Simon Mayer, Benedikt Ostermaie, "Moving application logic from the firmware to the cloud: towards the thin server architecture for the internet of things." In: 2012, 6th International Conference on Innovative Mobile and Internet Services in Ubiquitous Computing (IMIS), pp. 751–756. IEEE, July 2012.
24. Weiming Liu, Xueping Zhao, Jingfang Xiao, and Youlong Wu, "Automatic vehicle classification instrument based on multiple sensor information fusion." In: 3rd International Conference on Information Technology and Applications (ICITA 2005), vol. 1, pp. 379–382. IEEE, July 2005.
25. Peter Mell, Tim Grance, The NIST definition of cloud computing (2011)
26. Jiehan Zhou, Teemu Leppanen, Erkki Harjula, Mika Ylianttila, Timo Ojala, Chen Yu, Laurence Tianruo Yang, "Cloud things: A common architecture for integrating the internet of things with cloud computing." In: IEEE 17th International Conference on Computer Supported Cooperative Work in Design (CSCWD), pp. 651–657. IEEE, June 2013
27. Allan Cook, Michael Robinson, Mohamed Amine Ferrag, Leandros A. Maglaras, Ying He, Kevin Jones, Helge Janicke, "Internet of cloud: Security and privacy issues." In: Cloud Computing for Optimization: Foundations, Applications, and Challenges, Springer, Cham, 2018, pp. 271–301.
28. Jayavardhana Gubbi, Raja Kumar Buyya, Slaven Marusic, Marimuthu Palaniswami, Internet of things (IoT): A vision, architectural elements, and future directions. Future Gener. Comput. Syst. 29 (7), 1645–1660 (2013).

29. Armando Fox, Rean Griffith, Anthony Joseph, Randy Katz, Andrew Konwinski, Gunho Lee, David Patterson, Ariel Rabkin, and Ion Stoica, Above the clouds: a Berkeley view of cloud computing, vol. 4, pp. 506–522. Technical Report UCB/EECS-2009-28, EECS Department, University of California, Berkeley (2009)

30. Qi Jing, Athanasios V. Vasilakos, Jiafu Wan, Jingwei Lu, and Dechao Qiu, Security of the internet of things: Perspectives and challenges. Wirel. Netw. 20 (8), 2481–2501 (2014)

31. Md Tanzim Khorshed, ABM Shawkat Ali, and Saleh A. Wasimi, A survey on gaps, threat remediation challenges and some thoughts for proactive attack detection in cloud computing. Future Gener. Comput. Syst. 28(6), 833–851 (2012)

32 Sicari, Sabrina, Alessandra Rizzardi, Luigi Alfredo Grieco, and Alberto Coen-Porisini, Security, privacy and trust in internet of things: the road ahead. Comput. Netw. 76, 146–164 (2015)

33. Hui Suo, Jiafu Wan, Caifeng Zou, and Jianqi Liu, Security in the internet of things: a review. In: 2012 International Conference on Computer Science and Electronics Engineering (ICCSEE), vol. 3, pp. 648–651. IEEE, Mar 2012.

34. Guanglei Zhao, Xianping Si, Jingcheng Wang, Xiao Long, and Ting Hu, A novel mutual authentication scheme for internet of things. In: Proceedings of 2011 International Conference on Modelling, Identification and Control (ICMIC), pp. 563–566. IEEE, June 2011

35. Wen, Quangang, Xinzheng Dong, and Ronggao Zhang. : Application of dynamic variable cipher security certificate in internet of things. In: 2012 IEEE 2nd International Conference on Cloud Computing and Intelligent Systems (CCIS), vol. 3, pp. 1062–1066. IEEE, Oct 2012

36. Jun-Ya Lee, Wei-Cheng Lin, and Yu-Hung Huang. .: A lightweight authentication protocol for internet of things. In: 2014 International Symposium on Next-Generation Electronics (ISNE), pp. 1–2. IEEE, May 2014

37. Jie Lin, Wei Yu, Nan Zhang, Xinyu Yang, Hanlin Zhang, and Wei Zhao, A survey on internet of things: Architecture, enabling technologies, security and privacy, and applications. IEEE Internet Things J. 4 (5) (2017), 1125–1142.

38. Manuel Díaz, Cristian Martín, and Bartolomé Rubio. State-of-the-art, challenges, and open issues in the integration of Internet of things and cloud computing. J. Netw. Comput. Appl. 67 (2016), 99–117.

39. Dave Evans, The internet of things: how the next evolution of the internet is changing everything. CISCO White Paper 1 (2011), 1–11.

40. Flavio Bonomi, Rodolfo Milito, Jiang Zhu, and Sateesh Addepalli, Fog computing and its role in the internet of things. In: Proceedings of the First Edition of the MCC Workshop on Mobile Cloud Computing, pp. 13–16. ACM, Aug 2012.

11 Improved Privacy Preservation Framework for Cloud-Based Internet of Things

N. Yuvaraj
ICT Academy
Tamil Nadu, India

R. Arshath Raja
ICT Academy
Tamil Nadu, India

T. Karthikeyan
Annamacharya Institute of Technology and Sciences
Andhra Pradesh, India

N. V. Kousik
Galgotias University
Uttar Pradesh, India

CONTENTS

11.1 INTRODUCTION

In the field of telecommunication, the Internet of Things (IoT) comprising of objects embedded together with a communication channel is used for gathering and interchanging data. The number of devices and locations connected and the functionality they carry out are expected to increase in the coming years.

As there is a rapid growth in the number of connected devices, the amount of data will thus also increase. It will no longer be possible to store this data locally and temporarily. Rental storage space will be necessary. In addition, this enormous amount of data should be used as it should be. The data must be processed not only for the purposes of informing and developing knowledge but also to make the user a means of knowledge. This demands more processing that is not possible with low cost and lightweight devices at the IoT end. Again, there must also be processing and calculation on rent.

With cloud computing, all this is possible. The new paradigm for integration IoT and cloud computing is [1, 2]. IoT offers sophisticated means of communication through ubiquitous networks and devices with the internet.

Cloud computing, on the other hand, provides scalable access for the network in line with the requirements [3]. The data finally reaches the cloud, which, as required, stores, processes, and secures the data. The end-user residing on the other side of the cloud on the access layer shall be able to access the service once it is created.

It will not be so easy to allow all to become an IoT part and then have all the resources available via cloud computing. There are certain problems that need to be addressed in order for cloud-IoT to prevail. The cloud also has a business perspective, other than data and resources.

The homomorphic encryption [4] has high potential to secure IoT data by providing privacy in the user's IoT information, when the blockchain-based IoT is integrated with homomorphic encryption. Homographic encryption is an encryption mechanism that solves issues relating to security and privacy. It allows third-party service providers (such as cloud servers) to operate on certain kinds of encrypted data, while maintaining user privacy, without decrypting the encrypted data [5–16].

Cloud IoT creates more business opportunities, which makes it an attacker goal. In hybrid clouds, where private and public clouds are used by companies, security, and privacy, identity protection is becoming extremely important. Heterogeneous networks that support various types of data and services are involved in Cloud IoT. In line with their requirements, the network needs to have the flexibility to support all types of data with QoS support [17–26].

In this article, homomorphic encryption is used to securely authenticate the data from IoT devices to the cloud. Homomorphic encryption standard computes the cipher-text to transform the raw data to encrypted one, and at decryption, it matches with the results of operations. This algorithm acts as privacy-preserving outsourced storage and computation in cloud IoT model.

The outline of the chapter is presented below: Section 11.2 provides the proposed homomorphic encryption on cloud IoT model. Section 11.3 evaluates the entire work. Section 11.4 provides conclusions.

11.2 PROPOSED METHOD

The integrated cloud-IoT offers secure services with distributed services in cloud environment. The information is broadly distributed in various data center worldwide to improve the authentication process and data availability in all areas and businesses since all sectors are linked together via an integrated platform for instant and clever exchange and collection of information. The proposed method uses homomorphic encryption to improve the security of data transmitting between the cloud servers and IoT devices.

11.2.1 HOMOMORPHIC ENCRYPTION

Homomorphic encryption is a form of encoding which allows specific types of calculations to be performed on cipher texts and to obtain an encoded result that is the encoding text of the result of plain text operations. The implementation of the standard methods of encryption is dilemmatic: if the data is unencrypted, the storage/database service provider can be informed of sensitive information.

On the other hand, the provider cannot operate on it when it is encrypted. If data is encrypted, it is typically necessary to download and decrypt the entire database content to answer even a simple counting request.

Homomorphic encryption can be used by a user without first having to decrypt it. Homomorphic encryption is of considerable relevance to IoT networks since privacy can be maintained at all stages of the information communication, particularly without intermediate nodes.

For example, in middle nodules, data aggregation with operations such as amounts and averages can eliminate a lot of processing and storage. In turn, this reduces energy consumption, which is relevant to restricted environments. But note that such homomorphic cryptosystems are more computer-intensive and need longer control keys than standard symmetrical key algorithms to achieve a similar level of safety [27].

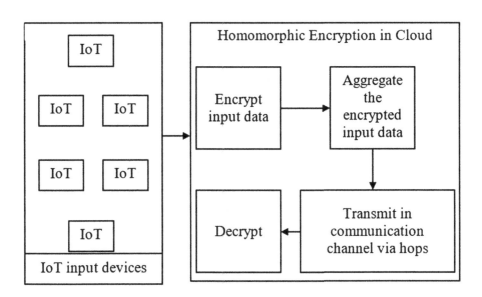

FIGURE 11.1 Architecture of homomorphic encryption in cloud IoT.

Generally, nodes are required to perform the following operations for secure homomorphic encryption mechanisms:

Step 1: Prior to the transmission to cloud, the data is encrypted using homomorphic encryption algorithm (E).

Step 2: Aggregate the data using an aggregation function.

Step 3: Encrypt the aggregate data prior to retransmission using E, and it is transmitted via hops.

Step 4: The received data packets from IoT devices are decrypted with inverse Homomorphic Encryption i.e. $D = E - 1$ for retrieving the original data at the receiver end.

There are four homomorphic encoding procedures, unlike public key encryption that has three security procedures: key generation, encryption and decryption. The homomorphic encryption allows service providers to carry out certain types of transactions with encrypted user data while maintaining the privacy of the encrypted user data without decrypting the encrypted data.

If the user wants to request some information on the cloud server in homomorphic encryption, he first encrypts the data and stores the cloud encrypted data. The user then sends query data to the cloud server after a while.

The cloud server uses homomorphic encryption to perform a prediction algorithm for encrypted data without being aware of the content of the data encrypted. Then the cloud will return the encrypted prediction to the user, and the user will decrypt the data he receives using the secret key of the user while retaining the privacy of IoT device data.

For homomorphic encryption, a plaintext mathematical operation is equivalent to a further cipher text operation. Consider a simple homomorphic procedure with the corresponding operation for the plaintext cipher text.

$$E(m_1) = E_1^e$$

$$E(m_2) = E_2^e$$

Then,
Addition homomorphism is defined as:

$$E(m_1) + E(m_2) = m_1^e + m_2^e$$

$$E(m_1) + E(m_2) = (m_1 + m_2)^e$$

$$E(m_1) + E(m_2) = E(m_1 + m_2)$$

Multiplication homomorphism is defined as:

$$E(m_1) \times E(m_2) = m_1^e \times m_2^e$$

$$E(m_1) \times E(m_2) = (m_1 \times m_2)^e$$

$$E(m_1) \times E(m_2) = E(m_1 \times m_2)$$

Consider P as the plaintext space with P lying with $\{0,1\}$, where the message tuple are embedded within $(m_1, m_2, ..., m_n)$. C is represented as the cloud and $C(m_1, m_2,..., m_n)$ as the ordinary function notation for the representation of cloud on the message tuple.

The pseudocode of HE is described below:

Step 1: Gen(1λ, α) is regarded as the key generation algorithm that helps in the generation of triplets of output keys that includes secret key-pair (sk, pk), evaluation key (evk), here λ is treated as the security parameter and α is regarded as the auxiliary input, thus (sk, pk, evk) $\leftarrow$ KeyGen($\$$)

Step 2: $Enc(p_k, m)$ defines the message (m) with the public key (p_k) and then the ciphertext ($c \in C$) is generated as output, where $c \leftarrow Enc_{pk}(m)$

Step 3: $Dec(s_k, c)$ is a decryption function that decrypts the cipher texts having a secret key (s_k), and then the original message (m) is received, where $m \leftarrow Dec_{sk}(c)$

Step 4: $Eval(evk, C, c_1, c_1, ..., c_n)$ is the output evaluated by considering the evk as the input key, a $C \in C$ and input cipher texts tuple is given as i.e., $c_1, c_2, ..., c_n$ and then the past results of evaluation is c* $Eval_{evk}(evk, C, c_1, c_1, ..., c_n)$.

11.3   RESULTS AND DISCUSSIONS

The performance of the proposed homomorphic encryption in cloud IoT is compared with existing cloud based homomorphic encryption. The validation is carried out against the size of cipher text during encryption, computational overhead, and finally, the execution time.

The results of cipher text key size for both encryption and decryption during homomorphic encryption in cloud IoT and cloud environment is given in Table 11.1–11.4.

TABLE 11.1

Encryption Key Size (in bytes)

Data Transmitted (kB)	Homomorphic Encryption in Cloud	Homomorphic Encryption in Cloud IoT
250	1568.16	1424.05
500	1910.42	1879.13
750	2383.52	2170.20
1000	2798.79	2639.51
1250	3363.86	3145.80
1500	3568.28	3482.19

TABLE 11.2

Decryption Key Size (in bytes)

Data transmitted (kB)	Homomorphic encryption in Cloud	Homomorphic encryption in Cloud IoT
250	1548.55	1406.24
500	1886.54	1855.64
750	2353.72	2143.073
1000	2763.80	2606.516
1250	3321.81	3106.478
1500	3523.67	3438.663

TABLE 11.3

Performance Evaluation on Computational Overhead (ms) for Encryption

Data Transmitted (kB)	Homomorphic Encryption in Cloud	Homomorphic Encryption in Cloud IoT
250	13.46	11.62
500	17.87	17.07
750	20.62	19.76
1000	22.64	21.04
1250	23.87	23.39
1500	24.26	23.85

TABLE 11.4

Performance Evaluation on Computational Overhead (ms) for Encryption

Data Transmitted (kB)	Homomorphic Encryption in Cloud	Homomorphic Encryption in Cloud IoT
250	13.29	11.47
500	17.65	16.86
750	20.36	19.51
1000	22.36	20.78
1250	23.57	23.10
1500	23.96	23.55

TABLE 11.5
Performance Evaluation on Execution Time (ms) for Encryption

Data Transmitted (kB)	Homomorphic Encryption in Cloud	Homomorphic Encryption in Cloud IoT
250	6.49	5.89
500	7.19	6.72
750	8.08	7.78
1000	8.52	8.14
1250	9.24	8.73
1500	10.15	9.25

The result shows that the proposed cloud IoT homomorphic encryption achieves reduced key size than the cloud homomorphic encryption. It further improves the security of the data in an effective manner such that no data breaches occur in this framework.

The results of computational overhead for both encryption and decryption during homomorphic encryption in cloud IoT and cloud environment are given in Table 11.2. The result shows that the proposed cloud IoT homomorphic encryption achieves reduced computational overhead than the cloud homomorphic encryption.

The results of execution time for both encryption and decryption during homomorphic encryption in cloud IoT and cloud environment is given in Table 11.3. The result shows that the proposed cloud IoT homomorphic encryption achieves reduced execution time than the cloud homomorphic encryption.

The results of Power Consumption for both encryption and decryption during homomorphic encryption in cloud IoT and cloud environment are given in Tables 11.4–11.8. The result shows that the proposed cloud IoT homomorphic encryption achieves reduced Power Consumption in the cloud than the cloud homomorphic encryption.

TABLE 11.6
Performance Evaluation on Execution Time (ms) for Decryption

Data Transmitted (kB)	Homomorphic Encryption in Cloud	Homomorphic Encryption in Cloud IoT
250	6.41	5.82
500	7.10	6.64
750	7.98	7.68
1000	8.41	8.04
1250	9.12	8.62
1500	10.02	9.13

TABLE 11.7

Average Power Consumption (J) for Encryption with 5 VMs

Data Transmitted (kB)	Homomorphic Encryption in Cloud	Homomorphic Encryption in Cloud IoT
250	43.52	52.87
500	44.58	54.16
750	46.82	56.88
1000	48.92	59.43
1250	51.23	62.23
1500	54.15	65.78

11.4 CONCLUSIONS

In this chapter, we designed a privacy-preserving model to improve the security of data in cloud IoT environment. The homomorphic encryption scheme is used to secure the IoT collected data and transmission of data along the cloud servers. This privacy-preserving encryption scheme increases the data security and makes it non-vulnerable to attacks like security breaches, improper updates, and bugs. The simulation results showed that the proposed homomorphic encryption in cloud IoT model offers improved security than the other methods.

TABLE 11.8

Average Power Consumption (J) for Decryption with 5 VMs

Data Transmitted (kB)	Homomorphic Encryption in Cloud	Homomorphic Encryption in Cloud IoT
250	42.98	52.21
500	44.02	53.48
750	46.23	56.17
1000	48.31	58.69
1250	50.59	61.45
1500	53.47	64.96

REFERENCES

1. Tian, Feng. "An Agri-Food Supply Chain Traceability System for China based on RFID & Blockchain Technology." In *2016 13th International Conference on Service Systems and Service Management (ICSSSM)*, pp. 1–6. IEEE, 2016.
2. Gentry, Craig. "Fully Homomorphic Encryption using Ideal Lattices." In *Proceedings of the Forty-First Annual ACM Symposium on Theory of Computing*, pp. 169–178. 2009.
3. Cheon, Jung Hee, Kyoohyung Han, Seong-Min Hong, Hyoun Jin Kim, Junsoo Kim, Suseong Kim, Hosung Seo, Hyungbo Shim, and Yongsoo Song. "Toward a Secure Drone System: Flying with Real-Time Homomorphic Authenticated Encryption." *IEEE access* 6 (2018): 24325
4. Armknecht, Frederik, Colin Boyd, Christopher Carr, Kristian Gjøsteen, Angela Jäschke, Christian A. Reuter, and Martin Strand. "A Guide to Fully Homomorphic Encryption." *IACR Cryptology ePrint Archive 2015* (2015): 1192.
5. Brakerski, Zvika, Craig Gentry, and Vinod Vaikuntanathan. "(Leveled) Fully Homomorphic Encryption without Bootstrapping." *ACM Transactions on Computation Theory (TOCT)* 6, no. 3 (2014): 1–36.
6. Tran, Julian, Farhad Farokhi, Michael Cantoni, and Iman Shames. "Implementing Homomorphic Encryption based Secure Feedback Control." *Control Engineering Practice* 97 (2020): 104350.
7. Min Zhao, E., and Yang Geng. "Homomorphic Encryption Technology for Cloud Computing." *Procedia Computer Science* 154 (2019): 73–83.
8. Alloghani, Mohamed, Mohammed M. Alani, Dhiya Al-Jumeily, Thar Baker, Jamila Mustafina, Abir Hussain, and Ahmed J. Aljaaf. "A Systematic Review on the Status and Progress of Homomorphic Encryption Technologies." *Journal of Information Security and Applications* 48 (2019): 102362.
9. Ullah, Shamsher, Xiang Yang Li, Muhammad Tanveer Hussain, and Zhang Lan. "Kernel Homomorphic Encryption Protocol." *Journal of Information Security and Applications* 48 (2019): 102366.
10. Alabdulatif, Abdulatif, Ibrahim Khalil, and Xun Yi. "Towards Secure Big Data Analytic for Cloud-Enabled Applications with Fully Homomorphic Encryption." *Journal of Parallel and Distributed Computing* 137 (2020): 192–204.
11. Chen, Xiu-Bo, Yi-Ru Sun, Gang Xu, and Yi-Xian Yang. "Quantum Homomorphic Encryption Scheme with Flexible Number of Evaluator based on (k, n)-Threshold Quantum State Sharing." *Information Sciences* 501 (2019): 172–181.
12. Yang, Jing, Mingyu Fan, and Guangwei Wang. "A Public Key Size Homomorphic Encryption Scheme based on the Sum of Sparse Subsets and Integers." *Cognitive Systems Research* 52 (2018): 543–549.
13. Xiong, Lizhi, and Danping Dong. "Reversible Data Hiding in Encrypted Images with Somewhat Homomorphic Encryption Based on Sorting Block-Level Prediction-Error Expansion." *Journal of Information Security and Applications* 47 (2019): 78–85.
14. Mainardi, Nicholas, Alessandro Barenghi, and Gerardo Pelosi. "Plaintext Recovery Attacks Against Linearly Decryptable Fully Homomorphic Encryption Schemes." *Computers & Security* 87 (2019): 101587.
15. Rahman, Mohammad Saidur, Ibrahim Khalil, Abdulatif Alabdulatif, and Xun Yi. "Privacy Preserving Service Selection using Fully Homomorphic Encryption Scheme on Untrusted Cloud Service Platform." *Knowledge-Based Systems* 180 (2019): 104–115.
16. Wazid, Mohammad, Ashok Kumar Das, Vivekananda Bhat, and Athanasios V. Vasilakos. "LAM-CIoT: Lightweight Authentication Mechanism in Cloud-based IoT Environment." *Journal of Network and Computer Applications* 150 (2020): 102496.

17. Razian, Mohammadreza, Mohammad Fathian, and Rajkumar Buyya. "ARC: Anomaly-Aware Robust Cloud-Integrated IoT Service Composition based on Uncertainty in Advertised Quality of Service Values." *Journal of Systems and Software* 164 (2020): 110557.

18. Serhani, M. Adel, Hadeel T. El-Kassabi, Khaled Shuaib, Alramzana N. Navaz, Boualem Benatallah, and Amine Beheshti. "Self-adapting Cloud Services Orchestration for Fulfilling Intensive Sensory Data-Driven IoT Workflows." *Future Generation Computer Systems* (2020).

19. Foughali, Karim, Karim Fathallah, and Ali Frihida. "A Cloud-IOT Based Decision Support System for Potato Pest Prevention." *Procedia Computer Science* 160 (2019): 616–623.

20. Ding, Li, Zhongsheng Wang, Xiaodong Wang, and Dong Wu. "Security Information Transmission Algorithms for IoT based on Cloud Computing." *Computer Communications* (2020), vol 155, pp.32–39.

21. Ding, Li, Zhongsheng Wang, Xiaodong Wang, and Dong Wu. "Security information transmission algorithms for IoT based on cloud computing." *Computer Communications* (2020).

22. Xu, Shengmin, Guomin Yang, Yi Mu, and Ximeng Liu. "A Secure IoT Cloud Storage System with Fine-Grained Access Control and Decryption Key Exposure Resistance." *Future Generation Computer Systems* 97 (2019): 284–294.

23. Wazid, Mohammad, Ashok Kumar Das, Rasheed Hussain, Giancarlo Succi, and Joel JPC Rodrigues. "Authentication in Cloud-Driven IoT-based Big Data Environment: Survey and Outlook." *Journal of Systems Architecture* 97 (2019): 185–196.

24. Kaur, Jasleen, and Pankaj Deep Kaur. "CE-GMS: A Cloud IoT-Enabled Grocery Management System." *Electronic Commerce Research and Applications* 28 (2018): 63–72.

25. Ray, Partha Pratim. "A Survey of IoT Cloud Platforms." *Future Computing and Informatics Journal* 1, no. 1–2 (2016): 35–46.

26. Karthikeyan, P., and M. Chandrasekaran. "Dynamic programming inspired virtual machine instances allocation in cloud computing." *Journal of Computational and Theoretical Nanoscience* 14, no. 1 (2017): 551–560.

27. Shirley, S. R., and P. Karthikeyan. "A survey on auction based resource allocation in cloud environment." *International Journal of Research in Computer Applications and Robotics* 1, no. 9 (2013): 96–102.

28. Cirani, Simone, Gianluigi Ferrari, and Luca Veltri. "Enforcing Security Mechanisms in the IP-based Internet of Things: An Algorithmic Overview." *Algorithms* 6, no. 2 (2013): 197–226.

12 IoT-Based Energy Management System

Design and Implementation and Its Security Challenges

M. Poongothai and N. Mahadevan
Coimbatore Institute Technology
Coimbatore, Tamil Nadu, India

CONTENTS

12.1 PREFACE

The electric energy consumption has been ever increasing, and it has resulted in the evolution of smart sensors to analyze and gain insights into power consumption patterns. The need for the proposed research is to automate and control physical devices in the laboratory/campus using data analyzed by the sensors, thereby reducing human intervention. It aims to ensure energy management by reducing the overall power consumption of the appliances with respect to the region occupied. It avoids unwanted driving of the load, thereby increasing the life span of devices [1].

> Enlightened Energy Management is not just about turning off unnecessary lights
> —it is about lightening the load on the electric grid using all our smartness.

Our campus has 40,000 sq m of built-up area for academic and administrative purposes. Out of which nearly 40–50% of space is occupied by laboratories. It has been identified that even appliances on standby mode consume up to 15% of electricity. So it is essential to automate all the electrical appliances of the laboratory to conserve power, which is wasted because of human negligence. Recent surveys have shown that 36% of the total energy produced is consumed by the educational institutions. To focus on energy and power management in buildings under this scenario is extremely important. In our campus so as to get better management of energy, the IoT-based smart energy management system is suggested. The main objectives of the proposed system are listed below:

- To design a proper system to get identification of occupancy and counting without human intervention.
- To predict user identity using machine learning algorithms based on the data collected from the real-time sensors and camera at the door.
- To detect the motion of the occupants using passive infrared sensor (PIR) thereby turning on/off the appliances only in respected area
- To sense the daylight intensity using LDR thereby turning on/off the appliances through the smart decisions.
- Thus complete automation of electrical appliances within a building can be achieved with the proposed research work.

The appliances can be monitored and controlled remotely, depending on the number of occupancies, thereby reducing the overall power of the building. The combined effects of all of these objectives integrated holistically into a smart laboratory, can reduce the energy consumption to about 20% and human intervention.

12.2 BACKGROUND AND RELATED WORK

In the present day scenario, energy management is one of the main problems that the globe faces, including developing countries like India. Fossil fuels are the main sources of energy supply, and hence they are in frequent use and depleting and hence increasing the energy rate. Therefore it is essential that serious steps are taken by commercial and industrial units to reduce the wastage of energy, reduce the cost of energy, and become energy efficient. In India, the main industries make use of 45percent of the 900 billion units of the power generated. There is a loss of 35 percent of the electric power which is generated, and these all losses are considered to be caused by the transmission and distribution (16 percent), theft (10 percent), among users some inefficiencies in energy (10 percent). The industrial and business-related users have 10 percent of inefficiencies especially among those using high KVA HT connections. Harmonic problems, fault in wiring, subsystem feedbacks and surrounding electrical systems are the main causes of inefficiencies. As a result of this, there is a drop in power factor and energy's higher utilization, which leads to high slab rates and penalties.

The smart home system and energy management system, which are existing have looked more into the electrical appliances control and electrical faults management in case of hazards. The energy management system has been developed in none of the study for the conservation of energy by the monitoring of the environmental surroundings and controlling the appliances accordingly.

Also, existing techniques like RFID, biometrics, and camera modules provide user identification with higher cost [2]. The existing model uses RFID tags for user identification that can be provided real-time response, but these tags must not be kept near metallic objects, and also the accuracy depends on the speed of walking. The height and weight sensors provide less accuracy and are not easily scalable.

Marinakis and Doukas (2018) [3] presented a modern and complex system based on IoT for the buildings and their intelligent energy management. Specifically speaking, the semantic framework for the combined and consistent modeling of each and every entity making up the smart building environments and their properties was suggested. On the suitable rules and general modus, it is aimed at the intelligent energy management for the smart buildings. A web-based tool was added in this work, and the main purpose is to enhance the energy management system of interactive buildings. In terms of energy of the buildings in real-time, the suggested tools collect, store, and represent the data. According to Shubham Mathur et al.(2017) [4], a human detector and counter with the help of Raspberry Pi was developed. It has been developed using the high level python programming language with an application of histogram oriented gradients descriptor. To detect the motion, the passive infrared motion sensor is used and the Raspberry Pi camera is enabled in case of motion detection condition. Ozturk et al. (2017) [5] suggested a solution for the demand response of the residential users. Adaptive neural fuzzy inference system (ANFIS) based system was created, which is used to predict the demands of energy as per the lifestyle of users and other energy consumption factors. The face identification system was introduced by Mohamed Heshmat et al. (2016) [6]. Variance estimation, facial feature extraction and CIE luv color space were introduced. It can be applied in the video surveillance, human-computer interfaces, management of image database, and smart home applications. Also, they have demonstrated the effectiveness of the proposed system in terms of various poses, expressions, and illumination conditions. The home energy management system on the basis of IoT is developed by Hlaing Thida Oo et al. (2016) [7], and this was developed for the Myanmar rural areas. The demand for electricity has been predicted in this work to satisfy the energy requirements. With the help of non-conventional energy sources such as solar and thermal, the energy demands could be met in this work.

Yi Liu et al., 2019, [8] proposed IoT-based energy management system for a smart city based on edge computing infrastructure with deep reinforcement learning. The performance of energy scheduling algorithms has been analyzed in terms of energy cost and delay with and without edge computing infrastructure. Wen-Tsai Sung and Sung-Jung Hsiaowe, 2020 [9], have addressed IoT Network security and applications via long-range technology, which was highlighting on multisensor fusion computing technology to implement a secure localization system of a wireless sensor network (WSN). LoRa IoT localization system security and sensor localization techniques have been analyzed for national security monitoring systems.

From the literature survey, it is inferred that any microcontroller platform and Internet of things (IoT) technology is suitable for automating all the electrical appliances in laboratories, homes, hospitals, buildings, and banks. This system also intimates the hazard of unauthorized entry and protection from fire accidents with immediate solutions. Machine learning algorithm based prediction mechanisms yields accurate result for individual appliance consumption forecasting. The pattern of the electricity consumption profile could be predicted. The existing energy management system developed by using microcontrollers, temperature, and LDR sensor, without taking the motion of the person entering the room and face recognition into account. This leads to unwanted driving of the load, even in undesired areas. Also, the occupancy status of the users was not provided in the existing systems. In order to provide a better enhancement to the existing system, the proposed model integrates with a machine learning algorithm to provide a real-time occupancy layout with respect to the region occupied by means of IoT [10].

12.3 IMPLEMENTATION OF IoT-BASED SMART ENERGY MANAGEMENT SYSTEM

With new emerging technologies having a greater impact on the lives of human beings, there is an urge to move on to a smarter technology that incorporates both energy management and reduction in human intervention. Taking a step towards this transition is acknowledging the interactions in occupant buildings, which often are helpful in the management of personalized energy and comfort making use of smart IoT. To offer real-time view of occupancy state, IoT-based smart energy management system has been suggested in this work with the help of smart sensors to reduce the intervention of humans.

12.3.1 PROPOSED SYSTEM DESIGN

Figure 12.1 shows the system design of the proposed system, which extracts data provided by ultrasonic sensors and pi cameras for aiding user identification in the

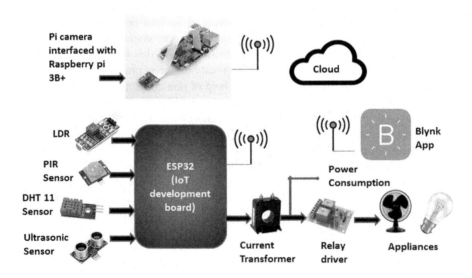

FIGURE 12.1   System model of the proposed system.

laboratory to improve security. The proposed smart energy management system consisting of a Raspberry pi 3B+, ESP32 Wi-Fi module, Pi-cameras, PIR sensors, ultrasonic sensors, light intensity sensors (LDR), and ambient temperature sensor. A door frame with ultrasonic sensors and pi camera is properly installed on a door. An ultrasonic sensor is placed at the top of the door to detect the entry of a person when he walks via the door. On striking the person, these are often reflected back as echo signals, and hence the target-oriented distance is computed, which is based on the time frame between the signal emission and echo reception. For ensuring secured entry as well as face recognition, Raspberry pi microcontroller with a pi camera is used. If either of the lasers is cut and then it can raise an exception to the Raspberry Pi, which in turn triggers the cameras to start capturing the images. The captured images are stored in the database, and with the help of the database, comparison can be made regarding the identification of persons entering the confined area. The laser-cut order represents the event's entry or exit. After the second laser is cut, the sensors are then reset.

PIR sensor senses the motion of the human and predicts whether he/she is within the confined range or not. With the help of PIR, occupancy detection and voltage switching are combined in a single package. This information is loaded into the IoT development board. Also, the LM35 temperature sensor measures the temperature status of the room and turns on the fan/cooler if needed. Similarly, the LDR calculates the luminous intensity of the room and turns on the lights in the desired location. Taking into account all these sensor data, the relay drivers are used to drive the appliances in the concerned region [11].

The Raspberry Pi assigns a unique session id to each entry or exit, and facial readings are recorded and tagged to this unique id, and the storage is made in the database. In the time of training, each event is tagged with the ground truth, such as information about the person with the help of installed Pi cameras at the doors. With respect to the motion detected by the PIR sensors, the occupancy layout is provided through the random forest classifier machine learning algorithms. Whenever a person is being captured by the Pi camera, the captured image undergoes a random forest classifier machine learning algorithm carried out by the Raspberry Pi module. If the data is matched, then the captured image is sent to the cloud, otherwise, an alert notification along with image is sent to the mobile application [12].

12.3.2 AUTOMATION OF ELECTRICAL APPLIANCES

Automation of electrical appliances is done by connecting current sensors and IR sensors to each appliance, and power consumption is measured using current sensors for future analysis.IR sensors are able to measure emitted heat by an object and detection of its motion. A short burst of ultrasonic sound is transferred to the target with the help of ultrasonic sensors, and sound is reflected back to the sensors by it. In this work, IR sensors are used to trigger fans and lights inside the lab/hall, and ultrasonic sensors HC-SR04 are used to count the entries of a particular lab/hall to disable unnecessary appliances. The main advantage of this work is that the occupancy-based lighting system is incorporated to manage/avoid power wastage. Apart from this, the current sensor is also added to measure the power consumed by each appliance to calculate the inimitability of the system. Taking into consideration, the optimal number of occupants and the light intensity measured by the LDRs, the appliances are turned on/off through the smart decisions taken by these sensors.

12.3.3 MEASUREMENTOF CURRENT CONSUMPTION USING CURRENT SENSORS

As a comparative study of installing indoor air quality systems and laboratory automation systems, current sensors are added to each and every appliance to measure power consumption, and detailed analysis was carried out with and without installing both the systems. The results revealed that with energy automation systems consume lesser power. In the proposed work, it is assumed that an office works for 5 hours duration, and the employee takes a break in between which improves his concentration and to get a feel of being relaxed. It is assumed that he took a break of 1½ hour, and thus, he totally works for 3½ hour duration. Our proposed system uses an occupancy-based laboratory automation system, and thus, appliances will be turned off whenever an employee leaves out for a break, and thus, the effective time with automation is chosen as 3.5 hrs and without automation is chosen as 5 hrs. The difference between the power consumption in both the case is said to be the total saving. Finally, the saving for the entire year was also calculated, assuming a year has 365 days [13]. The cost of one unit is considered to calculate the total cost saving throughout the year after installing an automation system inside the lab. Below is the mathematical calculation for the same:

If suppose, Load is 600W.
  Time Duration for calculating energy is 5hr.
  $E(kW/hr) = E(W) * T(hr)/1000(W/kW)$
  E (kW/hr) = Energy Consumed (kWhr)
  E (W) = Energy of load (Watt)
  T (hr) = Duration of load in ON state
  Then Energy Consumed without automation.
  $E = 600*5/1000 = 3$ (kWhr)
  When Automatic System is ON t_{eff}=3.5 hr then,
  $E = 600*3.5/1000$
  =2.1 (kWhr)
  Total saving =3 (kWhr) – 2.1 (kWhr)
  =0.9 (kWhr)
  If this pattern is for all day then total energy saving throughout the year=0.9*365
  =328.5 (kWhr)
  As we know 1 Unit = 1kWhr
  Therefore, total energy saved is 328.5 Units.
  Assuming cost of 1 Unit is 12Rs.
  Therefore, total cost saving through the year is = Rs. 328.5*12 = Rs. 3,942

12.4 EXPERIMENTAL RESULTS

The proposed smart energy management system has proved to be very efficient and effective. The designed prototype has no manual requirement of the ON or OFF switching in a case when a person moves in or out of the room. At the entrance of the room doors, the Pi cameras are placed with ultrasonic sensors in such a way that when a person moves in the room or leaves, the camera can capture the person. A machine learning algorithm is run in the image which is captured for the recognition of the person. The microcontroller circuit used is ESP32, for the lights and fans control in the room on the basis of PIR sensors, using LDR sensors and with the help of LM35 temperature. The record of entry and exit of a person is recorded with the help of it. The energy consumed before and after the system installation is tested with the help of this system.

Figure 12.2 shows the sensor board set up with three PIR sensors, two ultrasonic sensors, LDR, and an LM35 temperature sensor. Along with controller board set up with ESP32 microcontroller, relay module, and electrical appliances. A Raspberry Pi board Pi 3B+ is used as the controller board. Pi camera, which is deployed at the entrance, is used to capture the image of the user. The captured image information is used as training data for a machine learning algorithm. A Micro SD card is inserted into the slot on the board, which acts as the hard drive for the Raspberry Pi, which also holds the database. Whenever a person is being captured by the Pi camera, the captured image undergoes a machine learning algorithm carried out by the Raspberry Pi module.

The captured image is processed by a random forest classifier machine learning algorithm carried out by the Raspberry Pi module. This algorithm is able to predict the occupant's identity based on the captured image after training. The image

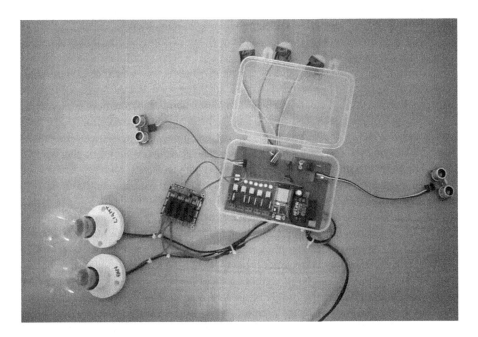

FIGURE 12.2   Hardware setup of the proposed energy management system.

captured by the pi camera is compared with the dataset and if matched, it triggers the pins high, thereby turning on the corresponding appliances.

Two ultrasonic sensors are installed at the doorstep along with the hardware module for the identification of the person entering/exiting the room.

This hardware setup has been implemented in our laboratory, which provides remote monitoring and control of the lab appliances using mobile applications thereby reducing power consumed and human energy considerably. This setup is installed in a very strategic location in the room so that the humans could be detected very effectively and efficiently.

ESP32 microcontroller serves to receive data from the cloud, and based on the processing results, it triggers the PIR sensor to start detecting for human motion. Using passive infrared technology, occupancy detection is combined by these sensors, and also they help in switching of the voltage in a single package. If a room is vacant for 5–10 minutes, these units turn off the lights automatically. Passive infrared technology is used by the ceiling mount sensors to detect the vacancy of the room, and automatically, they turn off the lights. The field of view is 180 and 360 degrees, and the total area they can cover up to 1000 square feet. PIR sensor setup has been installed in our laboratory to detect human motion. Based on sensor data, the appliances are controlled automatically

Figure 12.3 shows the Blynk application where the power consumption in the graphical representation can be viewed, and also the appliances can be turned ON/OFF manually using the switch buttons present in the Blynk application. The real-time data about the power consumption patterns can also be exported in CSV format, which can be used for future statistical analysis.

The user status saying that the person is inside the laboratory is indicated by the yellow lines, and the user leaving the lab is indicated by the red lines. Also, the current flow sensed by the current sensors, along with the power consumption, is visualized through the Blynk API. In addition to this, the appliances can be turned on manually if needed through the buttons provided in the Blynk home screen. This system also enables in turning on the appliances for the predefined timestamp, which is set via the mobile application. The user can set the time of availability inside the laboratory, and the appliances will be triggered only for the particular time interval and go off after the time limit is exceeded.

Blynk API offers a provision for the user to input their preferred temperature level either above or below the sensed information. Assuming that the sensed temperature level is 29°C since the user preferred temperature is 31°C which is slightly higher than the sensed room temperature there is no necessity to switch on the AC and the LED is in off state due to the room temperature is below the user preferred temperature. Assuming that the sensed temperature level is 37°C since the user preferred temperature is 31°C which is lower than the sensed room temperature, it is necessary to switch on the AC, and the LED is in on state due to the room temperature is above the user preferred temperature.

The Blynk App provides an option of exporting the output in CSV format where the real-time sensor data collected during various time stamps and the power consumption of appliances can be viewed in an excel file to be used for future statistical analysis to predict the energy profile of the lab/room.

The Blynk screen provides an option of exporting the output in CSV format, as shown in Table 12.1, where the real-time sensor data collected during various time stamps and the power consumption of appliances can be viewed in an excel file.

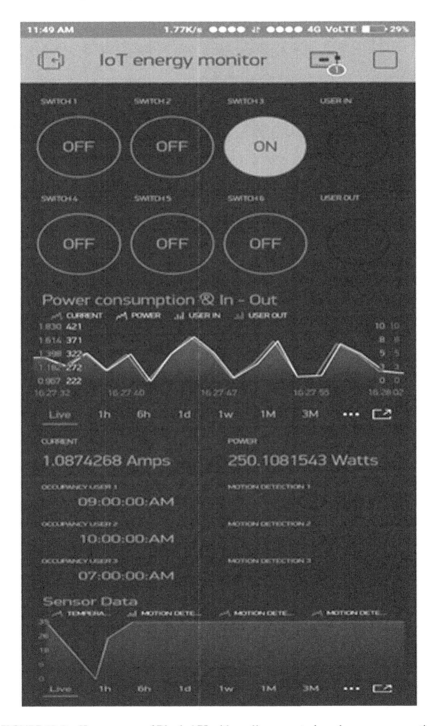

FIGURE 12.3 Home screen of Blynk API with appliance controls and power consumption estimation.

TABLE 12.1

Real-Time Sensor Data Collected and Power Consumption of Appliances during Various Time Stamps

S. No	Time stamp	Current (Amps)	Power (Watts)	Temperature (Celcius)	User in	User out	LDR output	Occupancy 1	Occupancy 2	Occupancy 3	Proximity 1	Proximity 2	Proximity 3
1.	11.18AM	0.37542	297.337	37	1023	0	1023	1023	1023	0	1023	0	1023
2.	11.19AM	1.91698	225.193	37	1023	0	1023	1023	1023	0	1023	0	1023
3.	11.20AM	1.95185	289.670	37	1023	0	1023	1023	1023	0	1023	0	1023
4.	11.21AM	1.72345	205.206	37	1023	0	1023	1023	1023	0	1023	0	1023
5.	11.22AM	1.60490	214.803	37	1023	0	1023	1023	1023	0	1023	0	0
6.	11.23AM	1.55869	219.539	36	1023	0	1023	1023	1023	0	1023	0	0
7.	11.24AM	2.91383	200.483	37	1023	0	1023	1023	1023	0	1023	0	0
8.	11.25AM	2.55994	280.279	37	1023	0	1023	1023	1023	0	1023	0	0
9.	11.26AM	1.39442	326.728	37	1023	0	1023	1023	1023	0	1023	0	0
10.	11.27AM	1.23932	301.453	36	1023	0	1023	1023	1023	0	1023	0	0
11.	11.28AM	1.19386	314.421	37	1023	0	1023	1023	1023	0	1023	0	0
12.	11.29AM	1.28477	294.138	37	1023	0	1023	1023	1023	0	1023	0	0
13.	11.30AM	1.13257	252.754	36	1023	0	1023	1023	1023	0	1023	0	0
14.	11.31AM	1.68292	254.361	36	1023	0	1023	1023	1023	0	1023	0	0
15.	11.32AM	1.31657	238.462	36	0	1023	1023	1023	1023	0	1023	0	0
16.	11.33AM	1.29438	291.660	36	0	1023	1023	1023	1023	0	1023	0	0
17.	11.34AM	0.97910	242.715	36	0	1023	0	1023	1023	0	1023	0	0
18.	11.35AM	1.25943	251.940	35	0	1023	0	1023	1023	0	1023	0	0
19.	11.36AM	0.89220	244.153	35	0	1023	0	1023	1023	0	1023	0	0
20.	11.37AM	0.93392	249.423	35	0	1023	0	0	1023	0	0	0	0
21.	11.38AM	0.95451	268.855	36	0	1023	0	0	1023	0	0	0	0
22.	11.39AM	0.87166	241.671	37	0	1023	0	0	1023	0	0	0	0

Blynk application has been developed, which gives the power consumption in the graphical representation, which can be viewed, and also the appliances can be turned ON/OFF manually. The real-time data about the power consumption patterns can also be exported in CSV format, which can be used for future statistical analysis.

12.5   SECURITY CHALLENGES

In this research, a smart energy management system is proposed to reduce the manual effort using IoT technology. The entire electrical appliances in the CIT campus are automated by connecting each and every appliance and laboratory projector with the network connectivity and sensors. Exchange of the real-time data, as well as physical object connection is enabled by the software [14].

The proposed system aims to collect and store the energy data of buildings such as energy consumption and weather data via Appliance control portal, energy usage patterns occupancy as well as predicted data produced by prediction models in real-time. The proposed architecture contains three modules, where each module is layered as a data sensing layer, network layer and application layer [15].

The data sensing layer contains smart IoT testbeds that are deployed in a building environment and are used to sense and gather data and react to specific conditions. The testbed is equipped with sensing devices and uses prediction models with machine learning to understand human activity and its impact on energy use of the building. The sensing layer involves two way communications between smart IoT testbed and gateway. For the communication between IoT testbed and gateway, the networking layer is held responsible and also among the cloud server. A bridge is there between cloud server and smart objects and is called a gateway module. For the management of applications, the application layer is responsible and also it analyzes collected data, generates the information and suggests important decisions. Each layer has its own security issues and challenges in the deployment of IoT in heterogeneous environments, which could affect the rapid development of IoT-based systems. There are several challenges and issues that need to be addressed before and during the design stages of IoT-based applications. The most important security issues and challenges faced on the IoT-based energy management system have been discussed in this section. Here we will discuss the possible security threats for each layer [16].

12.5.1   SECURITY ISSUES AND CHALLENGES IN DEPLOYMENT OF IOT-BASED ENERGY MANAGEMENT SYSTEM

Perception or physical layer: It consists of IoT devices such as Sensors, RFID readers, actuators, microcontrollers, and sensor gateway. The challenges are the physical security of sensors/devices, integrity of data transmission, and authentication, and access control of sensors/devices [17]. In IoT-based energy management system, sensors/devices are communication through public infrastructure; an intruder can easily access the energy consumption profile of laboratory/institute.

Network layer: It collects and processes the data from physical or perception layer. The challenges are access control of various networking devices, authentication between sensors and routers, and DoS attack. IoT devices will communicate with various heterogeneous devices either directly or through gateways. Heterogeneity requires security, which can be solved by implementing effective cryptographic algorithms to provide high throughput for end-to-end secure communication. In our system, an unauthorized person gets the necessary credentials and steal data from packets transmitted through IoT gateway in a smart energy building. Along with this, by the installed IoT sensors, the stolen information is collected, and malicious code can be induced by the hacker along with introducing viruses inside the networking traffic. This could then be sent to mobile applications and other systems.

The application Layer and data processing: the data analytics will be performed by this layer to help in the generation of pattern, and also to the end-users, the alert messages are generated. There is an authentic challenge here and in the app control and cloud storage, also in the privacy of data and its sensors. A device that is able to compromise threats in our system sometimes end up in a situation whereby sensors fail to detect some serious risks such as fire or flood, strange or suspicious movement inside the building [18].

12.5.2   RECOMMENDATIONS AND FURTHER WORK

The following recommendations have been suggested to address the above security issues and challenges in IoT-based heterogeneous environments [19]:

Authentication
Data integrity
User's privacy
Authorization and control access

When IoT is being designed, it is essential to keep in mind the security, minimization of data, transparency increase, especially among the end users. There should always be a choice for unexpected services in the system as well [20]. Along with this, systematically data security, as well as its protection, is supposed to be considered and addressed at the time of designing.

12.6   CONCLUSION

IoT-based energy management system was proposed where electrical appliances such as lights, fans, and air-conditioners are monitored and controlled based on the motion of occupants remotely through a Blynk application. The energy consumption of these appliances is measured by using a current sensor and hosted to cloud server in a timely manner, which is used to visualize the status of the appliances, whether it is ON/OFF in the graphical representation. It showed that the proposed system reduces power consumption and human intervention in monitoring the appliances to a greater scope in that way, supporting inefficient energy management in buildings. Up to 30 percentage of energy can be saved in our institute by employing IoT-based smart laboratory system on the whole in a year. The average cost of installing a smart system is $150 per room for fully automated labs with monitoring.

This chapter suggested only a brief review of security issues and challenges in the deployment of IoT-based energy management system, and the most important security services were recommended for the IoT-based heterogeneous environments.

REFERENCES

1. Fotopoulou, Eleni, Anastasios Zafeiropoulos, Fernando Terroso-Sáenz, Umutcan Şimşek, Aurora González-Vidal, George Tsiolis, Panagiotis Gouvas, Paris Liapis, Anna Fensel, and Antonio Skarmeta. "Providing personalized energy management and awareness services for energy efficiency in smart buildings." Sensors 17, no. 9 (2017): 2054.
2. Stergiou, Christos, Kostas E. Psannis, Byung-Gyu Kim, and Brij Gupta. "Secure integration of IoT and cloud computing." *Future Generation Computer* Systems 78 (2018): 964–975.
3. Marinakis, Vangelis, Haris Doukas, Evangelos Spiliotis, and Ilias Papastamatiou. "Decision support for intelligent energy management in buildings using the thermal comfort model." *International Journal of Computational Intelligence Systems* 10, no. 1 (2017): 882–893.
4. Mathur, Shubham, Balaji Subramanian, Sanyam Jain, Kajal Choudhary, and D. Rama Prabha. "Human Detector and Counter using Raspberry Pi Microcontroller." In 2017 Innovations in Power and Advanced Computing Technologies (i-PACT), pp. 1–7. IEEE, 2017.
5. Ozturk, Yusuf, Datchanamoorthy Senthilkumar, Sunil Kumar, and Gordon Lee. "An intelligent home energy management system to improve demand response." *IEEE Transactions on smart* Grid 4, no. 2 (2013): 694–701.
6. Heshmat, Mohamed, Walaa M. Abd-Elhafiez, Moheb Girgis, and Seham Elaw. "Face identification system in video." In 2016 11th International Conference on Computer Engineering & Systems (ICCES), pp. 147–154. IEEE, 2016.
7. Oo, Hlaing Thida, Khin Than Mya, Nyain Nyain Lwin, "IoT based home of energy management system for rural area in Myanmar", ICT Virtual Organization ASEAN Institutes and NICT ASEAN IVO Forum 3 (2016): 67–103.
8. Liu, Yi, Chao Yang, Li Jiang, Shengli Xie, and Yan Zhang. "Intelligent edge computing for IoT-based energy management in smart cities." IEEE Network 33, no. 2 (2019): 111–117.
9. Sung, Wen-Tsai, and Sung-Jung Hsiao. "IoT network security and applications via long range technology." Sensors and Materi.
10. Vermesan, Ovidiu, and Peter Friess, eds. Internet of things-from research and innovation to market deployment. Vol. 29. Aalborg: River publishers, 2014.
11. Karami, Majid, Gabrielle Viola McMorrow, and Liping Wang. "Continuous monitoring of indoor environmental quality using an Arduino-based data acquisition system." *Journal of Building* Engineering 19, (2018): 412–419.
12. Van Kranenburg, Rob. The internet of things: A critique of ambient technology and the all-seeing network of RFID. Institute of Network Cultures, 2008.
13. Khatoun, Rida, and Sherali Zeadally. "Cybersecurity and privacy solutions in smart cities." *IEEE Communications* Magazine 55, no. 3 (2017): 51–59.

14. Staudemeyer, Ralf C., Henrich C. Pöhls, and Bruce W. Watson. "Security and Privacy for the Internet of Things Communication in the Smart City." In Designing, Developing, and Facilitating Smart Cities, pp. 109–137. Springer, Cham, 2017.

15. Hameed, Sufian, Faraz Idris Khan, and Bilal Hameed. "Understanding security requirements and challenges in internet of things (IoT): A review." Journal of Computer Networks and Communications 2019.

16. Ali, Inayat, Sonia Sabir, and Zahid Ullah. "Internet of things security, device authentication and access control: A review." arXiv preprint arXiv:1901.07309 (2019).

17. Al-Fuqaha, Ala, Mohsen Guizani, Mehdi Mohammadi, Mohammed Aledhari, and Moussa Ayyash. "Internet of things: A survey on enabling technologies, protocols, and applications." IEEE Communications Surveys & Tutorials 17, no. 4 (2015): 2347–2376.

18. Sfar, Arbia Riahi, Enrico Natalizio, Yacine Challal, and Zied Chtourou. "A roadmap for security challenges in the internet of things." Digital Communications and Networks 4, no. 2 (2018): 118–137.

19. Khan, Nida Saddaf, Sayeed Ghani, and Sajjad Haider. "Real-time analysis of a sensor's data for automated decision making in an IoT-based smart home." Sensors 18, no. 6 (2018): 1711.

20. Kouicem, Djamel Eddine, Abdelmadjid Bouabdallah, and Hicham Lakhlef. "Internet of things security: A top-down survey." Computer Networks 141 (2018): 199–221.

13 MQTT

As Default, Secured Protocol for IoT Communication and Its Practical Implementation

S. Suresh Kumar
QIS College of Engineering and Technology
Ongole, Andhra Pradesh, India

D. Palanivel Rajan
CMR Engineering College
Hyderabad, Telangana, India

Yogesh M. Iggalore
METI M2M India Private Limited
Mysore, Karnataka, India

CONTENTS

13.1 INTRODUCTION

Before getting into MQTT (message queuing telemetry transport) protocol and its practical implementation, let us understand what IoT is all about. IoT stands for Internet of Things (IoT), which means physical things that are connected with each other or to a group of others either through the internet or intranet. For example, let us consider that a temperature sensor, a humidity sensor, a PIR sensor, a bulb, and a fan are connected to home Wi-Fi router[1]. PIR sensor broadcasts information about the presence of a human in the room to all other devices. Temperature and humidity sensors broadcast temperature, and humidity values to all other devices. Based on PIR sensor signal, temperature and humidity value: bulb and fan present in the room will be decided to be switched ON or OFF. This may look like automation, but actually, it is not. The example which was discussed above is about how well we use the resources available to us by conserving energy. It is not mere automation.

13.1.1 DEFINITION OF IoT

IoT is a system of interrelated computing devices, mechanical and digital machines, objects, animals, or people that are provided with unique identifiers and the ability to transfer data over a network without requiring human-to-human or human-to-computer interaction[2]. Figure 13.1 shows the goal of IoT and Figure 13.2 shows its benefit to humanity.

We can achieve a benefit to humanity by the following:

- Collecting data from all physical devices.
- Converting this raw data into information.
- Using that information, we need to gain knowledge.
- With that knowledge, we need to attain wisdom.

By the end of 2023, 1.3 billion IoT devices will be connected to the internet. To handle these huge numbers of devices, you need a solid, simple IoT protocol. To address these issues, so many IoT protocols are created, namely MQTT, CoAP, DDS, AMQP, Sigfox, and so on. In this chapter, we will look into the MQTT protocol only.

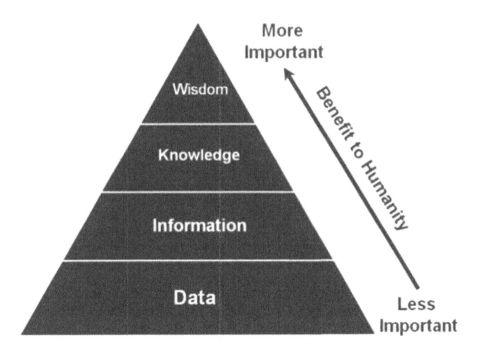

FIGURE 13.1   Goal of IoT.

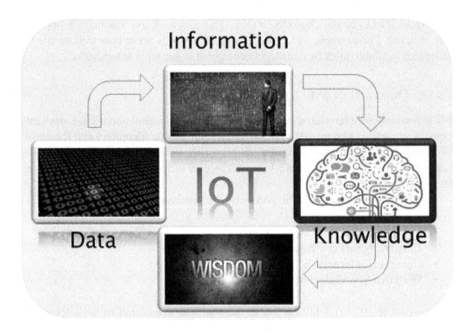

FIGURE 13.2   Benefit of IoT to humanity.

13.2 MQTT

MQTT stands for message queuing telemetry transport protocol. In the year 1999, IBM employee Andy and Eurotech employee Nipper developed the first version of MQTT for monitoring and controlling the oil pipeline in desert. As HTTP protocol is heavy for small hardware devices, and also the devices are connected via satellite link, the requirement of this monitoring hardware is that it should use low bandwidth, low power, and it should be lightweight, hence MQTT. In the year 2013, IBM made MQTT open source.

13.2.1 Definition of MQTT

MQTT is a client-server publish/subscribe messaging transport protocol. It is lightweight, open, simple, and designed for easy implementation. These characteristics make it able to ideal for use in many situations, including constrained environments, such as for communication in machine-to-machine (M2M) and IoT contexts where a small code footprint is required and/or network bandwidth is at a premium.

As it is mentioned in the definition, MQTT is a client-server publish/subscribe messaging transport protocol, which means there will be a centralized server for all MQTT clients which handle all connections and communications. There can be multiple MQTT clients connected to a single server. As shown in above Figure 13.3, a temperature sensor, home devices, IoT devices, phone, and laptop are connected to a centralized MQTT server.

Message Queuing Telemetry Transport

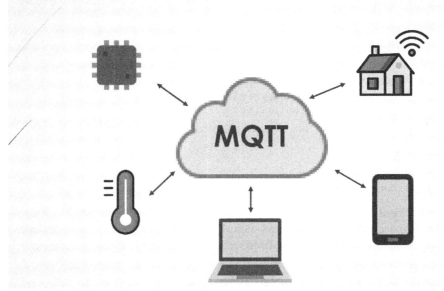

FIGURE 13.3   Interface of MQTT.

13.2.2 Pub/Sub Architecture

MQTT works on Pub-/Sub architecture, which means publish and subscribe architecture. This is different from traditional client-server architecture, as shown in Figure 13.4. In client-server architecture client directly talks to the end server whereas in MQTT publisher and subscriber never contact each other directly[3].

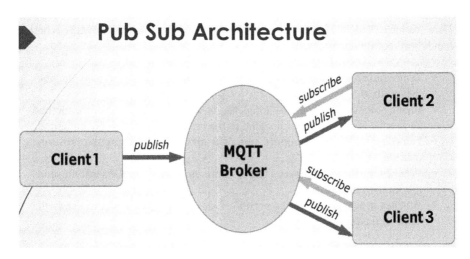

FIGURE 13.4   Pub/Sub architecture.

13.2.3 Benefits of Pub/Submodel

1. Client devices need not remember all other clients they need to communicate to.
2. It helps resource constraint devices to communicate with each other.
3. No restriction in number of client connections.
4. All communications are central, hence less overhead for clients.
5. All communications work under path called "topics".
6. No exchange of IP addresses of publishers and subscribers is required.
7. Time decoupling can be avoided.
8. Synchronization is not required since publish and subscribe are independent actions.

13.2.4 Important Terminology used In Pub/Sub Architecture

1. **Client**: these are called end devices in MQTT or simply MQTT client devices. Both publishers and subscribers are MQTT clients. These devices communicate with a centralized server called broker[4]. Clients can communicate with other clients indirectly by subscribing to a topic or by publishing a topic to broker.
2. **Broker:** It is the brain of the MQTT protocol where all the connections and communications happen. It is responsible for

   i. Establishing a connection with client devices.
   ii. Registering to publish and subscribe to topics.
   iii. Responsible for receiving all the client messages.
   iv. Filtering of received messages.
   v. Receiving published messages from the client.
   vi. Sending publish messages to subscribed client.

Here we have 3 MQTT clients and centralized MQTT broker. Here MQTT broker handles all communications.

- Temperature sensor sends its temperature value to the broker using the publish method.
- Phone requests the broker for temperature value using the subscribe method and sends fan on/off control signal to the broker using the publish method.
- Fan requests for on/off control signals from the broker using the subscribe method.
- Whenever a broker receives a temperature message from the sensor it sends the same to the phone.
- Whenever a broker receives an on/off signal from the phone it sends the same to the fan.

Here all clients don't know each other but work together with publish-subscribe model.

Any number of these devices can be connected to the broker depending on broker handling capacity.

13.2.5 MQTT Client Broker Communication

MQTT protocol works on top of TCP/IP protocol. Here the broker will never initiate the connection. The connection is always initiated by the client. The broker acknowledges the connection request. Once the connection is established, the client can publish as many messages as it wants, and every publishes massage will be acknowledged by the broker[5, 6, 7, 8, 9]. If the client is interested in any messages, then it can subscribe to that message, and the broker sends back subscribe acknowledgement. Whenever the broker receives a related subscribe message from any other client, it will forward the same to this client[10]. The client can disconnect from the broker, whenever it wants, and it will receive disconnect acknowledgement. (This feature is only available in MQTT 5 but not in MQTT 3).

13.2.6 MQTT Standard Header Packet

MQTT works with exchanging series of packets[11].
Each packet contains three sections:

1. Fixed header (Which is always present in all MQTT packets)
2. Variable header
3. Payload

Variable header and payload are not always present in all the packets.

13.2.6.1 Fixed Header

Each packet contains a fixed header. The first byte in the fixed header is the control packet[12, 13]. In this, the last four bits are allocated for control packet type, and the first four bits are allocated for control packet flags. Depending on the variable header and payload, packet length varies from one to four bytes.

13.2.6.2 Variable Header and Payload

Since these two packets are dynamic, we will discuss them later.

13.2.7 Connect Packet

Connect packet contains a fixed header, variable header, and payload. Once the TCP/IP connection is established, the client will send connect packet as its first packet to the broker[14].

13.2.7.1 Fixed Header

In the fixed header, the first byte is a connect packet with a value of 0x10. Here no control packets. Values of bytes 2, 3, 4, and 5 depend on variable header and payload.

13.2.7.2 Variable Header

The variable header for connect packet contains four fields in the following order Protocol name, protocol level, connect flags, and keep alive.

Protocol name: first 2 bytes contain protocol name length with a value of $0 \times$ 04 next 4 bytes are protocol name as "MQTT".

Protocol level: Byte7 is protocol level, value is 0×04 which means MQTT version 3.1.1.

Connect flags: connect flag byte contains a number of parameters specifying the behaviour of the MQTT connection. They also indicate the presence and absence of their specific fields in the payload.

Connect flags are given as below:

Username: if the username flag is set to 0, it must not be present in the payload.

Password: if the password flag is set to 0, it must not be present in the payload.

Will retain, Will QoS (quality of service)**, and Will flag:** these flags decide the will message to publish to another client when this client disconnects ungracefully. If **will flag** is set to 1, then **will message** should be present in the payload. For example, consider temperature sensor sets it **will message** as "temperature sensor in room is offline", then if the temperature sensor is not sending its values on regular intervals, the broker will transmit **will message** of temperature to all its subscribers (clients).

Situations in which will message gets published include for example, and many more

1. An I/O error or network failure detected by the MQTT broker.
2. The client fails to communicate within the Keep Alive time.
3. The client closes the network connection without first sending a DISCONNECT packet.
4. The MQTT broker closes the network connection because of a protocol error.

If the **Will Flag** is set to 1, the **Will QoS** and **Will Retain** fields in the Connect Flags will be used by the MQTT broker, and the **Will Topic** and **Will Message** fields MUST be present in the payload.

The Will Message must be removed from the stored Session state in the MQTT broker once it has been published, or the MQTT broker has received a disconnect packet from the client.

If the **Will Flag** is set to 0, the **Will QoS** and **Will Retain** fields in the Connect Flags MUST be set to zero, and the **Will Topic** and **Will Message** fields must not be present in the payload.

If the Will Flag is set to 0, a Will Message must not be published when this Network Connection ends

Keep alive: 2 bytes of keep alive value in seconds define the longest period of time that the broker and client can endure without sending a message. Within this time, client must send a ping packet to say that it is alive. Max time can this value can be set is 18 hours 20 minutes.

Clean session: if clean session flag is set to 0, the broker must resume the previous session connection, otherwise it must discard the previous session connection.

Session state in client consists of **QoS** and the existence of session, which will be discussed later in this chapter.

13.2.7.3 Payload

Connect packet payload contains the following fields Client id, Last will topic, Last will message, username, and password; these are based on the flags set in connect flags. The client id is an identification name given to IoT devices, which can be the same or unique[15]. Usually, unique ID is preferred. If the client sets will topic flag, the payload must include its Last will topic, and Last will message. Username and password field must be included based on their respective flags

1. First 2 bytes will be client id length and followed by client id.
2. Followed by 2 bytes of will topic length and will topic.
3. Followed by 2 bytes of will message length and will message.
4. Followed by 2 bytes of username length and username.
5. Followed by 2 bytes of password length and password.

13.2.8 CONNACK PACKET

The CONNACK packet is sent by the broker in response to a connect packet received from the client[16]. It is a connection acknowledgement packet from broker to the client. Fate of further communication depends on this. This packet includes an only fixed header and variable header. The payload is not included here.

13.2.8.1 Fixed Header

The first byte of a fixed header is connection acknowledge packet with a value of 0x20. The second byte is the length of remaining bytes with a value of 0x02, which means 2 bytes will be followed by this.

13.2.8.2 Variable Header

The variable header includes 2 bytes of connection flags and connection return code.

If the broker accepts the connection with a clean session set to 1, then the broker must send the response with the flag set to 0. If the broker accepts connection with the clean session set to 0, then the broker must send the response with a flag set based on the previous session connection.

The value of connection return code indicates the reason for connection failure or success. Following are only possible return codes from the broker:

1. Value 0×00 connection accepted.
2. Value 0×01 connection refused, unacceptable protocol version.
3. Value 0×02 connection refused, identifier rejected.
4. Value 0×03 connection refused, server error
5. Value 0×04 connection refused, bad username or password
6. Value 0×05 connection refused, unauthorized

Let us understand the above mentioned (CONNECT and CONNACK) packets with the following example.

Connect packet from client to broker

13.2.8.2.1 Connect Packet Example

The fixed header of connect packet value is 0x10, and it has 0x50 bytes as remaining bytes. The variable header includes protocol length, name, level and keep alive value. Connect flags with **QoS 1** and **will flag** and **retain flag** set to 1. Keep Alive: 0x00 0x3C (60 Seconds) server-client timeout here is 60 seconds if no communication between client and broker, the broker will disconnect with client after 90 seconds which is one and half of keep alive time. max time from the client side is 18:12:15 hours, and for sever, it is 27:24:30 hours (one and half of 18:12:15).

13.2.8.2.2 Payload

For connect packet, the payload order is client id, will topic, will message, username, and password. In this example, client id length is 17 bytes, and client id is *CC:50:E3:9B:F7:84* (mac address of device). Will topic length is 24 bytes and will topic is (*CC:50:E3:9B:F7:84/status*). Will message length is 7 bytes, and message is *offline*. Username length is 6 bytes, and username is **Yogesh**. Password length is 6 bytes and password is **Yogesh**. In order to respond back, MQTT broker sends CONNACK packet. Let us now look into it.

CONNACK packet is simple, and it has only 4 bytes. Byte1 is connect packet identifier with a value 0×20. Byte2 is the remaining length of the packet. Here it is 0x02. Byte3 is a connect flag with value 0. It indicates that the connection success and it is a clean session. Byte4 is return code value 0x00 indicates connection accepted.

13.2.9 PUBLISH PACKET

MQTT client can publish messages as soon as it is connected to the broker. MQTT works on the topic-based filtering of the incoming messages[17]. Each publish message must include a topic so that the broker can filter it out and forward it to an interested client. Each message should have payload, which contains the data to transmit in byte format. MQTT is data agnostic, the client can send binary data, text data, or even full-fledged XML or JSON. Publish message has several attributes. Let us discuss them in detail.

13.2.9.1 Topic Name

The word **topic** refers to a UTF-8 string that the broker uses to filter messages for each connected client[18]. It includes one or more topic level. Each topic level is separated by a forward slash.

Topics are case sensitive which means "Home/room" is different from "home/room". Space should not be used in the topic. A topic uses a wildcard (using + and #) as well; single level and multi-level.

Topics beginning with $symbol have a different purpose. These are not part of the subscription when you subscribe to multi-level wildcards. $symbol reserved for internal statistics purposes.

Best practices to use topics are

1. Never use a leading forward slash since it introduces unnecessary topic levels with zero characters.
2. Never use space in the topic.
3. Keep topic names short and concise.
4. Use only ASCII characters, avoid non-printable characters.
5. Embed a unique identifier or the client id into the topic.
6. Don't subscribe to multi-level wild cards (#) as we do not know how many topics are present
7. Don't forget extensibility while choosing the topic name.
8. Use a specific topic, not a general one.

13.2.9.2 Quality of Service

QoS is a key feature of MQTT. QoS gives the client the power to choose a level of service that matches its network reliability and application logic[19]. It is basically an agreement between the sender of the message and receiver of the message that defines the guarantee of delivery for the specific message. There are 3 QoS levels in MQTT

1. At most once (0)
2. At least once (1)
3. Exactly once (3)

While we talk about QoS in MQTT, we need to consider two things. One: message received by broker from publishing client and Two: message received by subscribing client from the broker. We will look QoS separately; the broker transmits the message to subscribing client with QoS either equal to or less than that of publishing client.

13.2.9.2.1 QoS 0—At Most Once

This is the minimal QoS and fastest in MQTT. This service level guarantees a best-effort delivery, but there is no guaranty in delivery. No acknowledgment will be sent by the broker or subscribing client. By default, all publishing messages will have QoS 0 unless mentioned otherwise. QoS is fire and forgets and guarantees the delivery of messages depending on TCP protocol.

13.2.9.2.2 QoS 1—At Least Once

This is the second level of QoS provided by MQTT. QoS 1 provides a guarantee that message gets delivered at least once to the receiver. There can also be a possibility of receiving the same message multiple times. The sender stores the message until the receiver acknowledges. The sender uses the packet identifier in each packet to match the publish packet with the corresponding PUBACK packet. If the sender does not receive a PUBACK packet in a reasonable amount of time, the sender resends the PUBLISH packet. When a receiver gets a message with QoS 1, it can process it immediately. For example, if the receiver is a broker, the broker sends the message to all subscribing clients and then replies with a PUBACK packet.

If the publishing client sends the message again, it sets a duplicate (DUP) flag. In QoS 1, this DUP flag is only used for internal purposes and is not processed by the broker or client. The receiver of the message sends a PUBACK, regardless of the DUP flag.

13.2.9.2.3 QoS 2—Exactly Once

QoS 2 is the highest level of service from MQTT. This level guarantees 100% that each message is received only once by intended recipients. QoS 2 is the safest and slowest QoS level. The guarantee is provided by at least two request/response flows, A four-part handshake. The sender and receiver use the packet identifier of the original publish a message to coordinate the delivery of the message.

> Use QoS 0 when clients have stable internet connection, and messages are not critical.
> Use QoS 1 when the client needs to send all the messages to the broker, fast and stable service.
> Use QoS 2 when messages are critical and the client don't want his receiver to miss any messages or duplications.

13.2.9.3 Retain Flag

This flag decides whether the message is saved in the broker as the **known good value** for a specific topic. When a new client subscribes to a topic, this saved message will be transmitted by the broker to a newly subscribed client.

13.2.9.4 DUP Flag

This flag indicates the message which is already sent, and it is duplicate.

Now let us look into PUBLISH packet. Publish packet includes a fixed header, variable header, and payload.

13.2.9.5 Payload

This is the actual content of the message. MQTT is data-agnostic. It is possible to send images, text in any encoding, encrypted data, and virtually every data in binary.

13.2.9.6 Packet Identifier

The packet identifier uniquely identifies a message as it flows between the client and the broker. The packet identifier is only relevant for QoS levels greater than zero. The client library and/or the broker is responsible for setting this internal MQTT identifier.

13.2.9.7 Fixed Header

Fixed header, in that first byte, has PUBLISH packet code in higher nibble and publish flags in lower nibble. While publishing a message to the broker, the client must set its QoS, duplicate flag, and retain flag as required by the client. Byte 2 to byte 4 will decide the length of the packet.

13.2.9.8 Variable Header and Payload

First 2 bytes give the length of the topic followed by topic and at packet identifier at the end (this is present only if QoS in the fixed header set QoS 1 or QoS 2). The payload contains the message to be sent to the broker by the client or vice versa.

Example 1:

topic name "CC:50:E3:9B:F7:84/hall" payload "test"

Publish topic with QoS0

In this example the payload message test is publishing with QoS 0. The first byte is publishing command with QoS set to 0 and no duplicate flag. Retain flag set to 1.

Publish topic with QoS1

In this example the payload message test is publishing with QoS 1. The first byte is publishing command with QoS set to 1 and no duplicate flag. Retain flag set to 1.

Publish topic with Qos2

In this example the payload message *test* is publishing with QoS2. The first byte is publishing command with QoS set to 2 and no duplicate flag. Retain flag set to 1.

13.2.10 PUBACK PACKET

This packet is a response to the publish packet sent with QoS 1. It includes a fixed header and variable header but not payload.

The first byte is a publish acknowledge packet and value is 0×40, the second byte is the length of remaining bytes it is always 2, and third and fourth bytes are packet identifier used for QoS 1.

Example 2:

PUBACK contains 4 bytes. The first byte is **PUBACK** packet code, value 0x40 followed by the length of remaining bytes that is 0×02 and at last 2 bytes of packet identifier. **PUBACK** packet has no payload.

13.2.11 PUBREC Packet

Publish received is the first response to the publish packet of QoS 2. It includes a fixed header and variable header but not payload.

The first byte is a publish receive packet, and value is 0×50, the second byte is the length of remaining bytes it is always 2, and third and fourth bytes are packet identifier used for QoS2.

> **Example 3:**
>
> **PUBREC** contains 4 bytes. First byte **PUBREC** packet code, value 0×50 followed by length of remaining bytes that is 0×02 and at last 2 bytes of packet identifier. **PUBREC** packet has no payload.

13.2.12 PUBREL Packet

Publish release packet is the response to the publish received packet. This is the third packet in QoS 2. It includes a fixed header and variable header but not payload.

The first byte is a publish release packet, and value is 0x60, the second byte is the length of remaining bytes, it is always 2, and third and fourth bytes are packet identifier used for QoS2.

> **Example 4:**
>
> **PUBREL** contains 4 bytes. First byte **PUBREL** packet code, value 0x60 followed by length of remaining bytes that is 0×02 and at last 2 bytes of packet identifier. **PUBREL** packet has no payload.

13.2.13 PUBCOMP Packet

Publish complete packet is the response to the publish release packet. This is the fourth and last packet in QoS 2. It includes a fixed header and variable header but not payload.

The first byte is a publish complete packet and value is 0x70, the second byte is the length of remaining bytes it is always 2, and third and fourth bytes are packet identifier used for QoS2.

> **Example 5:**
>
> **PUBCOMP** contains 4 bytes. First byte **PUBCOMP** packet code, value 0x70 followed by length of remaining bytes that is 0×02 and at last 2 bytes of packet identifier. **PUBCOMP** packet has no payload.

13.2.14 Subscribe Packet

Publishing messages do not make sense if no one ever receives it. It means that no client has subscribed to the topic of the message. To receive messages on topics of interest, the client sends a subscription message to the MQTT broker. This subscribe message is very simple, it contains a unique packet identifier and list of subscriptions. Subscribe packet includes a fixed header, variable header, and payload.

13.2.14.1 Fixed Header

First byte of fixed header is a subscribe packet code and value is 0x82.

Byte 2 to 5 denotes the packet length and varies depending on packet size.

13.2.14.2 Variable Header

A variable header will have 2 bytes of unique packet identifier.

13.2.14.3 Payload

The first 2 bytes of the payload will be the length of topic/s and followed by topic/s. The last byte decides QoS in which the client is interested.

Example 6:

Subscribe Topic is CC:50:E3:9B:F7:84/led

QoS 0:

In this example client subscribes to topic *CC:50:E3:9B:F7/led* with QoS 0 and packet identifier 0x00 0x01

QoS 1:

In this example client subscribes to topic *CC:50:E3:9B:F7/led* with QoS 1 and packet identifier 0x00 0x01.

QoS 2:

In this example client subscribes to topic *CC:50:E3:9B:F7/led* with QoS 1 and packet identifier 0x00 0x01.

13.2.15 SUBACK Packet

Subscribe acknowledge packet is the response to the subscribe packet. Based on QoS requested in subscribe packet, the content in **SUBACK** packet varies. **SUBACK** packet includes fixed header, variable header and payload.

13.2.15.1 Fixed Header

First byte of fixed header is a subscribe acknowledge packet code and its value 0x90 second byte is remaining length of the packet and it is 0x03.

13.2.15.2 Variable Header

Variable header includes 2 bytes of packet identifier.

13.2.15.3 Payload

SUBACK packet includes single byte as payload. Based on QoS requested, the broker sends QoS, which is applicable to a particular subscribe request. If the response is 0x80 it means the subscribe packet is failed to register in the broker. Client must pay attention when it receives response 0x80.

> **Example 7:**
>
> **SUBACK** contains 5 bytes. First byte **SUBACK** packet code, value 0x90 followed by length of remaining bytes that is 0x03 and followed by 2 bytes of packet identifier and last QoS supported.

13.2.16 UNSUBSCRIBE PACKET

The client can unsubscribe to a topic in the same way in which it subscribes. The unsubscribe packet will help the client in achieving the same. It includes a fixed header, variable header, and payload.

13.2.16.1 Fixed Header

The first byte of a fixed header is a unsubscribe packet code, and value is 0xA2.
Byte 2 to 5 are packet length and vary depending on packet size.

13.2.16.2 Variable Header

A variable header will have 2 bytes of unique packet identifiers.

13.2.16.3 Payload

The first 2 bytes of the payload will be the length of topic/s and followed by topic/s.

> **Example 8:**
>
> Unsubscribe topic: CC:50:E3:9B:F7:84/led
>   Unsubscribe packet includes packet identifier as variable header and topic name as payload.

13.2.17 UNSUBACK PACKET

Unsubscribe acknowledge packet is the response to the unsubscribe packet. This packet has a fixed header and variable header and no payload.

13.2.17.1 Fixed Header

The first byte of a fixed header is an unsubscribe acknowledge packet code, and its value 0xB0 second byte is the remaining length of the packet and is 0x02.

13.2.17.2 Variable Header

A variable header includes 2 bytes of packet identifier.

Example 9:

Unsuback packet is a four-byte packet, which includes packet identifier and has no payload.

13.2.18 PINGREQ Packet

Ping request packet is used when client has no message to send to the broker but needs to inform the broker that client is alive[20]. On configured intervals, ping request is sent to the broker, and the broker will respond to it. Ping request packet includes only a fixed header packet.

13.2.19 PINGRES Packet

Ping response packet is a response packet from the broker to the client when the client sends **PINGREQ** packet. Ping request packet includes only a fixed header packet.

13.2.20 Disconnect Packet

Disconnect packet is the last packet sent by the client to the broker in order to disconnect from MQTT protocol.

13.3 SERVER-SIDE IMPLEMENTATION

Eclipse mosquitto is an open-source message broker that implements the MQTT protocol. It supports MQTT version 5, 3.1.1 and 3.1. Mosquitto is lightweight and is suitable for use on all devices from low single-board computers to full servers. The MQTT protocol provides a lightweight method of carrying out messaging using a publish/subscribe model[2]. This makes it suitable for IoT messaging, such as with low power sensors or mobile devices such as phones, embedded computers, or microcontrollers.

The mosquitto project also provides a C library for implementing MQTT clients, and the very popular mosquitto_pub and mosquitto_sub command line MQTT clients. Mosquitto is part of the Eclipse Foundation, is an iot.eclipse.org project, and is sponsored by cedalo.com.

Here mosquitto broker installation procedure is given below:

Step 1
Require a Linux server
.320
sudo apt-get update
sudo apt-get install mosquitto
using these two commands mosquitto broker is installed
Step 2
install mqtt clients
sudo apt-get install mosquitto-clients
Mosquitto clients help us easily test MQTT through a command line utility
Subscribe test in server
mosquitto_sub -t "test"
Publish test in server
Open another ssh server connection and publish a message in mosquitto broker
    with the command
mosquitto_pub -m "message from mosquitto_pub client" -t "test"
Step 3
Secure with the password
sudo mosquitto_passwd -c/etc/mosquitto/passwd dave
after this command type the password
Create a configuration file for Mosquitto pointing to the password file we have
    just created.
sudo nano/etc/mosquitto/conf.d/default.conf
This will open an empty file. Paste the following into it.
allow_anonymous false
password_file/etc/mosquitto/passwd
Save and exit the text editor with "Ctrl+O", "Enter" and "Ctrl+X".
Now restart Mosquitto server and test our changes.
sudo systemctl restart mosquitto
In the subscribe client window, press "Ctrl+C" to exit the subscribe client and
    restart it with following command.
mosquitto_sub -t "test" -u "dave" -P "password"
Note the capital -P here.
In the publish client window, try to publish a message without a password.
mosquitto_pub -t "test" -m "message from mosquitto_pub client"
The message will be rejected with following error message.
Connection Refused: not authorised.
Error: The connection was refused.
Now publish a message with the username and password.
mosquitto_pub -t "test" -m "message from mosquitto_pub client" -u "dave"
    -P "password"
Step 4
To add extra user
mosquitto_passwd -b passwordfile username password

use this command for adding an extra user

Note: Some servers don't allow direct port forwarding. So, it is suggested to allow a firewall for a particular port of MQTT in the server. Default port for MQTT 1883.

REFERENCES

1. https://www.codeinstitute.net/blog/what-is-iot/
2. https://www.iso.org/standard/69466.html
3. https://en.wikipedia.org/wiki/Publish%E2%80%93subscribe_pattern
4. https://www.hivemq.com/mqtt-essentials/
5. http://mqtt.org/documentation
6. http://docs.oasis-open.org/mqtt/mqtt/v3.1.1/os/mqtt-v3.1.1-os.html
7. https://thingsboard.io/docs/reference/mqtt-api/
8. https://docs.devicehive.com/docs/mqtt-api-reference
9. https://en.wikipedia.org/wiki/MQTT
10. https://www.thethingsnetwork.org/docs/applications/mqtt/
11. *MQTT Version 3.1.1.* Edited by Andrew Banks and Rahul Gupta. 29 October 2014. OASIS Standard. http://docs.oasis-open.org/mqtt/mqtt/v3.1.1/os/mqtt-v3.1.1-os.html. Latest version: http://docs.oasis-open.org/mqtt/mqtt/v3.1.1/mqtt-v3.1.1.html.
12. http://www.steves-internet-guide.com/mqtt-protocol-messages-overview/
13. https://docs.aws.amazon.com/freertos/latest/lib-ref/c-sdk/mqtt/structIotMqttNetwork Info__t.html
14. https://www.hivemq.com/blog/mqtt-essentials-part-3-client-broker-connection-estab-lishment/
15. https://docs.devicewise.com/Content/Products/IoT_Portal_API_Reference_Guide/ MQTT_Interface/MQTT-data-usage.htm
16. https://pubsubclient.knolleary.net/api.html
17. https://doc-snapshots.qt.io/qtmqtt/index.html
18. https://www.ibm.com/support/knowledgecenter/SSQP8H/iot/platform/reference/mqtt/ restrictions_limitations.html
19. https://docs.solace.com/MQTT-311-Prtl-Conformance-Spec/MQTT_311_Prtl_ Conformance_Spec.htm
20. https://www.espruino.com/MQTT
21. https://pypi.org/project/paho-mqtt/

Index